UNITED STATES NAVY HELICOPTER PATCHES

HELICOPTERS

COMMANDS

SCHOOLS

WINGS

SQUADRONS

MICHAEL L. ROBERTS

Schiffer Publishing Ltd
4880 Lower Valley Road • Atglen, PA 19310

Other Schiffer Books on Related Subjects:

United States Navy and Marine Corps Aviation Squadron Lineage, Insignia, and History,
ISBN: 978-0-7643-4755-9

Library of Congress Control Number: 2015955422

Designed by Michael L. Roberts and Justin Watkinson
Cover design by Justin Watkinson
Type set in Univers LT Std/Minion Pro

ISBN: 978-0-7643-5012-2
Printed in The United States of America

Published by Schiffer Publishing, Ltd.
4880 Lower Valley Road
Atglen, PA 19310
Phone: (610) 593-1777; Fax: (610) 593-2002
E-mail: Info@schifferbooks.com

For our complete selection of fine books on this and related subjects, please visit our website at www.schifferbooks.com. You may also write for a free catalog.

This book may be purchased from the publisher. Please try your bookstore first.

We are always looking for people to write books on new and related subjects. If you have an idea for a book, please contact us at proposals@schifferbooks.com.

Schiffer Publishing's titles are available at special discounts for bulk purchases for sales promotions or premiums. Special editions, including personalized covers, corporate imprints, and excerpts can be created in large quantities for special needs. For more information, contact the publisher.

DISCLAIMER: The patches depicted in this book are from the author's personal collection and are presented for historical purposes and to serve as a guide for historians, collectors, and designers. This book is not intended to be the complete authority on this subject. It depicts over 1,900 helicopter patches from the 1950s to the present. The helicopter photos in this book and on the cover are official US Navy photographs. All patches read from top to bottom on the page.

ACKNOWLEDGMENTS

The following individuals are honorably mentioned for their encouragement and support for this project:

US Navy
Hard Assets
Aces High Historical
Bamboo Tools J.L
Heritage Online LLC
Carl Altevogt
Doug Barnard
Daniel G. Benjamin
Robert Biondi
Pilippe Dognan
Robert Bottie
Robert Dugger
Johnny W. Evans
Michael Falls
Ursulina Housh
John M. Miller
Xavier Rodriguez Jimenez
Kara Jones
Dong Min Rhie
CDR J. E. Rimmer HSL-49 C.O.
Jay Roberts
Ian Robertson
Jocelyn Saverance
James R. Townley
LTJG Thomas J. Wendt USN

PHOTO CREDITS

US Navy
Dreamsamerica
Heritage Online Auctions
Istockhistory
Memorabiliarus
Newportgamesandphotoslic
Pretty Planes
SN Travis Alston USN
Cdr. Jane Campbell USN
PH1 Erecson USN
PHC Lawrence B. Foster
MCS2 Alan Gragg USN
JO2 Pete Hatzacos
MCS1 Joan E. Jennings USN
Bruce Lazarius
MCC Edward G. Martens
Carolyn & James Toomey
James R. Townley
Jeff Wilson
Sgt. Harold Woodrum USMC
Bradl Zarlin

CONTENTS

FOREWORD

rich heritage of artwork that has been used to adorn
y squadrons. Borrowing from business and sports, the
both a brand and a mascot. It can be spotted on uniforms,
ignage, coffee cups, and web sites. In previous books, Mr.
hat few, if any, other authors have done, which is provide
used by Naval aviation squadrons.

rticular book special is the ephemeral nature of patches
ments. In many US Navy helicopter squadrons, deployments
squadron of ten or more helicopters, instead by detachments
th twenty to twenty-five pilots, aircrew, and maintenance
ments (dets) have traditionally been given the privilege of
scot or theme and designing their own patch. With funds
ls, the orders for patches are relatively small and once the
stood down, the patch is retired. Few dets last beyond one
ment exercises to the end of deployment. Only the serious
can keep up with the hundreds of dets over the years and
es made in small quantities. The collection of so many images
tters to squadrons and detachments soliciting patches before

stated in this forward are only mine, I know that any sailor
ted with helicopter squadrons would appreciate this book that
ne with a passion for preserving the history of an old tradition
ng of naval aviation.

son E. Rimmer, US Navy
ficer, HSL-49

HU-1 Det. "D" Piasecki HUP-3, Buno. 147616, Tail code UP, Modex 33. USS *Coral Sea* (CVA-43) December 20, 1961.

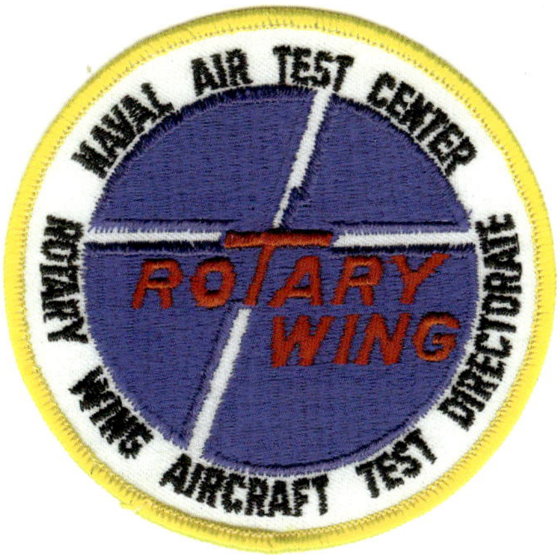

Naval Helicopter Association

Rotary Wing Aircraft Test Directive NATC Patuxent River, MD

Helicopter Capital NAS Imperial Beach, CA

Navy Helicopter Association

Rotary Wing Center of Excellence Naval Air Depot Pensacola, FL

Helicopter Capital OLF Imperial Beach, CA

Dynamic Interface NATC Patuxent River, MD

Med.-Evacuation Team Naval Hospital Camp Lejeune, NC

NAS Mayport, FL, Aircraft Intermediate Maintenance Department

NAS Mayport, FL

Helicopter Attack NAS Fallon, NV

Search and Rescue NAS Oceana, VA

NS Mayport, FL

Search and Rescue NAS Fallon, NV

Search and Rescue NAS Oceana, VA

Penguin Mk2 Mod 7 Missile

Search and Rescue NAS Lemoore, CA

Search and Rescue NAS Pensacola, FL

Search and Rescue NAS Whidbey Island, WA

HSC Weapons School Atlantic (Desert)

Helicopter Sea Combat Weapons School Pacific (Desert)

Search and Rescue NAS Whidbey Island, WA

Helicopter Sea Combat Weapons School Pacific

Rotary Wing Weapons School Fallon, NV

HSC Weapons School Atlantic

Helicopter Sea Combat Weapons School Pacific (Subdued)

Helicopter Maritime Strike Weapons School Pacific Fleet

Helicopter Maritime Strike Weapons School Pacific Fleet (Subdued)

Helicopter Maritime Strike Weapons School Pacific Fleet (Desert)

Helicopter Multi-Mission H-60R, H-60S Fleet Introduction Team

SH-60 Seahawk Flight Test team

Commander CHTW-1 CH-53E Super Team

Commander Helicopter Wing Reserve

Commander Helicopter Wing Reserve Det. 1 HS-84 and HS-85 HMCS Provider 1998

Commander ASW Wing Pacific NAS Cubi Point, Philippines

Commander HS Wing Atlantic

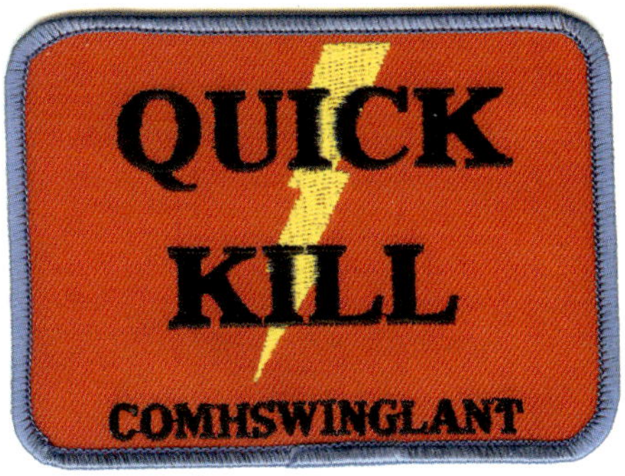

Commander HS Wing Atlantic Quick Kill

Helicopter Sea Combat Wing Atlantic

Commander HSL Wing Pacific Fleet

Commander HS Wing Pacific

Helicopter Sea Combat Wing Atlantic (Desert)

Commander Helicopter Sea Control Wing 1

Commander HS Wing 1

Commander HSL Wing Atlantic Fleet

Commander Helicopter Sea Control Wing 3

Commander Helicopter Tactical Wing Atlantic

SH-2F Seasprite Lamps MK 1

SH-3 Sea King

Commander Helicopter Tactical Wing Pacific

SH-2F Seasprite Maintenance

SH-3H Sea King

Commander Helicopter Tactical Wing 1

H-3 Alumni

SH-3H Sea King

SH-3H Search-Detect-Destroy

H-46 Phrog Endangered Species

H-46 Seaknight

Sea King 25 Years 1963–1988

H-46 Phrog Subdued

H-46 Service to the Fleet

H-46 1000 Hours

H-46 Sea Knight FRS Instructor

H-53 100 Hours FAM

MH-53E Sea Dragon

MH-53E FCF Required

MH-53E

MH-53E Sea Dragon

MH-53E FCF Pilot, That Others May Fly

MH-53E

MH-53E Sea Dragon (Subdued)

MH-53E MCM & Transport JMSDF

H-60 Armed Helicopter Firehawk

H-60 Armed Helicopter Strikehawk

H-60 Darts ASW

H-60 Lamps

H-60 Armed Helicopter HH-60H Integration

H-60 Armed Helicopter AN AAS-44 Flir/LRD

H-60 Lamps (Subdued)

SH-60 Armed Helicopter HSL Sea Strike

SH-60 Lamps

H-60 Lamps MK III

H-60 Seahawk

SH-60B 2000 Hours (Subdued)

SH-60B Lamps MK III 1000 Hours

H-60 Support CSAR

SH-60B 3000 Hours (Subdued)

SH-60B Lamps MK III 2000 Hours

SH-60B 1000 Hours (Subdued)

SH-60B Armed Helicopter Dirty Deeds Done Dirt Cheap

SH-60B Plane Captain

SH-60B Seahawk

SH-60F Seahawk Helicopter

SH-60F Seahawk CV Helicopter

SH-60B Seahawk (Subdued)

SH-60F Seahawk CV Helicopter

SH-60F, HH-60H CV Helicopter

SH-60B Seahawk

SH-60F Seahawk CV Helicopter

SH-60F, HH-60H Seahawk

SH-60F, HH-60H Seahawk (Subdued)

MH-60R Seahawk

SH-60R Seahawk

SH-60F, HH-60H Seahawk (Desert)

MH-60R Seahawk

MH-60S Knighthawk

Combat Search and Rescue

MH-60R Seahawk

SH-60S Knighthawk

MH-60S Knighthawk (Subdued)

HAL-3, VN

HAL-3 Born for Battle, VN

MH-60S Knighthawk (Desert)

HAL-3 Seawolf, VN

Flotilla 5 & HAL-3 Binh Thuy, VN

MH-60S 100 Hours, Centurion, Night Vision Goggles

HAL-3 Vung Tau, VN

HAL-3 Det. 2 Rung Sat, VN

HAL-3 Det. 3 Armed Helicopters VN

HAL-3 Det. 9, VN

HAL-4

HAL-3 Det. 5 Chien Dau Co, VN

HAL-4

HAL-4 (Subdued)

HAL-3 Det. 5 Filthy Five-Flying Death, VN

HAL-4 Red Wolves

HAL-5 Blue Hawks

HAL-5

HAL-5

HC-1 Det. Delta USS *Coral Sea* (CVA-43) Westpac 1964–1965

HAL-5

HAL-5 (Subdued)

HC-1 Fleet Angel Rescue

HAL-5 (Subdued)

HC-1 Fleet Angels

HC-1 Apollo 15 Mission

HC-1 Rowell's Rats

HU-1 HC-1, 1948-1994

HC-1 Det. Skylab

HC-1 Det. 2 Rescue

HC-1 Point Mugu Det.
January 1–February 03, 1989

HC-1 Det. 2 Desert Storm 1991

HAL-3 Sea Wolf Bell H-1 Huey over Vietnam.

HC-1 Det. 3 Wheeling

HC-1 Det. 4

HC-1 Det. 8 1969–1970

HC-1 Det. 3 Sultans of SAR

HC-1 Det. 4 Westpac 1977

HC-1 Det. 27 Armed Helicopter, VN

HC-1 Det. 3 USS *Tripoli* (LPH-10) Operation Desert Shield 1990–1991

HC-1 Det. 6

HC-2 Fleet Angels (Day)

HC-2 Fleet Angels (Night)

HC-2 Circuit Riders

HC-2

HC-2 Fleet Angels (Subdued)

HC-2 Fleet Angels

HC-2 Timmy's Angels

HC-2 Fleet Angels (Desert)

HC-2 Executive Transport

HC-2 Chicks Dig Rotorheads

HC-2 Chicks Dig Rotorheads

HC-2 Det. 1 Naples, Italy

HC-2 Det. 2 Desert Duck Airlines

HC-2 USS *Shangri-La* (CVA-38)
Med. Cruise 1969

HC-2 Det. 1

HC-2 Det. 2 Desert Duck Airlines

HC-2 Det. 1 Commander Sixth Fleet

HC-2 Det. 1 R.I.P. 1987–2005 R.I.P.

HC-2 Det. 2 Desert Duck Airlines (Desert)

HC-2 Det. 2 Desert Duck Airlines 1976–2005 (Sundown Patch)

HC-2 Det. 4 City of New York 2000

HC-2 Det. 62 USS *Independence* (CVA-62) Westpac 1965

HC-2 Det. 3 USS *Mount Whitney* (LCC-20) Commander Second Fleet

HC-2 Det. 11 USS *Intrepid* (CVS-11), VN 1966

HC-2 Det. 62 USS *Independence* (CVA-62) NATO Cruise 1969

HC-2 Det. 3 Nordik Raiders JTFX, Operation Strong Resolve 2002

HC-2 Det. 60 USS *Saratoga* (CVA-60) Med. 1969

HC-2 Det. 66 USS *America* (CVA-66) Westpac 1972

HC-2 Det. 67 USS *John F. Kennedy* (CVA-67) Med. Cruise 1970–1971

HC-3 Twenty-Two Years, 140,000 Hours Mishap Free

HC-3 Plane Captain

HC-3

HC-3 Twenty-Five Years Mishap Free

HC-3 Det. 104–287 Missions over Japan 1975

HC-3 Pack Rats

HC-3

HC-3 Det. 111 USS *Kiska* (AE-35) Westpac 1976–1977

HC-3 Det. 111 Westpac 1976–1977

HC-4 Plank Owner

HC-4 MH-53E Plane Captain, Sigonella, Sicily, Italy

HC-4 Combat Support

HC-4 Combat Support

HC-4 H-53 Heavy Combat Support

HC-4 Thundercats (Desert)

HC-4 Black Stallions (Desert)

HC-4-2003

HC-4-2003

HC-4 Wave One Chief 2006

HC-4 Det. 83 USS *Austin* (LPD-4) 1970-1971

HC-4 Sea Dragon

HC-4 Det. 1 Midnight Riders

HC-5 Providers

HC-4 Det. 2 RAF Akrotiri, Cypress

HC-5 Providers (Subdued)

HC-5 Providers

HC-5 Det. 2 Always in the Dark

HC-5 Det. 2

HC-5 Island Search & Rescue

HC-5 Det. 2 Humping 2 Please

HC-5 Det. 3 Knight Riders

HC-5 Det. 1 USNS *Kilauea* (T-AE-26), Explosive Ordnance Delivery

HC-5 Det. 2 USNS San Jose (T-AFS-7)

HC-5 Det. 4 Club Bohica 1990

HC-5 Det. 4 Operation Iraqi Freedom February 27, 2003-TBD

HC-5 Det. 6 Iron Gator Search & Rescue

HC-5 Det. 6 USS *Belleau Wood* (LHA-3)

HC-5 Det. 5 Saddam Busters

HC-5 Det. 6 Be Flexible

HC-5 Det. 6 Search & Rescue Dogs

HC-5 Det. 6 Last with Wood, First with *Essex*

HC-5 Det. 6 Devil Dog Search & Rescue

HC-5 Det. 6 Whatever it Takes

HC-6 Combat Support

HC-6 Det. 1 Search & Rescue

HC-6 Det. 1 USS *Mt. Baker* (AE-34) Med. 1996

HC-6 Logistics, Amphibious SAR, 30 Years of Excellence SEP. 1967

HC-6 Det. 1 USNS *Sirius* (T-AFS-8) Med. 1986

HC-6 Det. 1 USS *Inchon* (LPH-12) Operation Blue Harrier 1997

HC-6 Gray Dog Air Taxi USS *Kalamazoo* (AOR-6) Operation Desert Storm December 28, 1990–April 18, 1991

HC-6 Det. 1 USS *Santa Barbara* (AE-28) Med. December 1990-July 1991

HC-6 Det. 1 USS *Saipan* (LHA-2) SAR & NSW 1998

HC-6 Det. 2 USNS *Saturn* (T-AFS-10) July 20–November 5, 1987

HC-8 Det. 2 Med. Cruise 1997

HC-6 Det. 2 Double Deuce

HC-6 Det. 2 USS *Mt. Baker* (AE-34) Med.-Red Sea-Indian Ocean September 1991–March 1992

HC-6 Det. 2 USS *Arctic* (AOE-8) April–October 2001

HC-6 Det. 3 USS *Concord* (AFS-5) Costa Del Sol 1987–1988

HC-6 Det. 2 USNS *Saturn* (T-AFS-10) Med. 1995–1996

HC-6 Det. 2 Search & Rescue

HC-6 Det. 3 USNS *Saturn* (T-AFS-10) & USNS *Sirious* (T-AFS-8) February–September 1989

HC-6 Det. 3 Med. 1991

HC-6 Det. 4 USS *Guam* (LPH-9) SAR Det.

HC-6 Det. 4 USNS *Sirius* (T-AFS-8) January–June 2004

HC-6 Det. 3 USS *Guam* (LPH-9) Med. 1997–1998

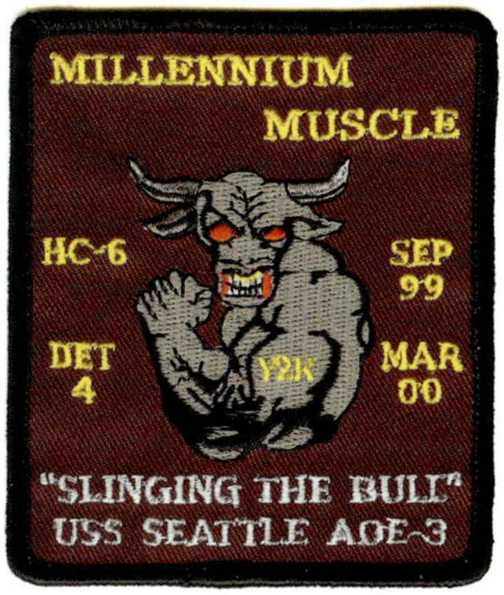

HC-6 Det. 4 USS *Seattle* (AOE-4) September 1999–March 2000

HC-6 Det. 4

HC-6 Det. 4 USNS *Saturn* (T-AFS-10) & USS *Concord* (AFS-5) Med. Cruise December 1993–May 1994

HC-6 Det. 4 USS *Seattle* AOE-4 Nomads 2002

HC-6 Det. 5 USS *Concord* (AFS-5) Med. 1985

HC-6 Det. 5 USS *Savannah* (CL-42) Med. 1086-1987

HC-6 Det. 5 USNS *K-200* Med., Red Sea, Gulf of Oman, Persian Gulf December 20, 1990–Victory

HC-6 Det. 5 USNS *Saturn* (T-AFS-10) & USNS *Concord* (T-AFS-5) Med. 1993

Bell TH-1L Huey Buno. 157806.

HC-6 Det. 5 USS *Kearsarge* (LHD-3) (Maiden Voyage) Med. 1995

HC-6 Det. 5 USNS *Mt. Baker* (T-AE-34) September 2002–June 2003 (Desert)

HC-6 Det. 6 USNS *Sirius* (T-AFS-8) May–October 1997

HC-6 Det. 5 USNS *Supply* (T-AOE-6) Med. 2000

HC-6 Det. 5 Operation Iraqi Freedom

HC-6 Det. 6 USNS *Saturn* (T-AFS-10) & USNS *San Diego* (T-AFS-6) Med. Cruise 1995

HC-6 Det. 5 USNS *Mt. Baker* (T-AE-34) September 2002–June 2003

HC-6 Det. 5 Operation Iraqi Freedom (Desert)

HC-6 Det. 6 USS *Nassau* (LHA-4) SAR 7 NSW Med. 1996–1997

HC-6 Det. 6 USS *Nassau* (LHA-4)

HC-6 Det. 6 USNS *Supply* (AOE-6) Devilhawks (Desert)

HC-6 Det. 7 USS *Butte* (AE-27) 1993–1994

HC-6 Det. 6 USS *Detroit* (AOE-4) Med. & Gulf Cruise 1998–1999

HC-6 Det. 7 USNS *Supply* (AOE-6) 1st Med. Cruise 1996

HC-6 Det. 10 USS *Concord* (AFS-5) & *Sirius* 1994–1995

HC-6 Det. 6 USNS *Supply* (AOE-6) Devilhawks

HC-6 Det. 7 USS *Santa Barbara* (AE-28) February–July 1998

HC-7

HC-7 Det. Cubi Point 1973 Champs. Cubi Point Captains Cup & Subic Bay Admirals Cup

HC-8 Dragon Whales

HC-8 Battle Readiness "E" 1992

HC-8

HC-8 Dragon Whales (Subdued)

HC-8 Battle Readiness "E" 1994

HC-8 (Subdued)

HC-8 Bay Raiders

HC-8 Battle Readiness "E" 1996

HC-8-20 Years of Excellence 1984–2004

HC-8 Det. 2 USS *Kearsarge* (LHD-3) Med. 1997

HC-8 Helicopter Det. Team 3 USNS *Sirious* (T-AFS-8) Spain, Italy, France, Africa 1989–1990

HC-8 Det. 1 USNS *Supply* (T-AOE-6) June–December 2002

HC-8 Det. 2 Last of the Breed, Sunset Patch

HC-8 Helicopter Det. Team 3 Hurricane Hugo Puerto Rico & U.S. Virgin Is. September–October 1989

HC-8 Team 2

HC-8 Det. 2 USS *Kearsarge* (LHD-3) ESQ. H-60 Black Sheep

HC-8 Det. 3 USNS *Mt. Baker* (T-AE-34) Med. & Gulf 2000

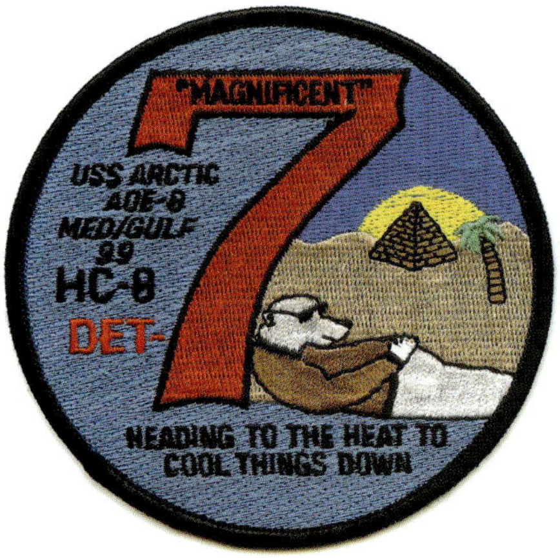

HC-8 Det. 3 USNS *Sirius* (T-AFS-8) 2003

HC-8 Team 5 Outlaws

HC-8 Det. 7 USNS *Sirius* (T-AFS-8) Arabian Gulf February–July 1996

HC-8 Det. 4 USS *Detroit* (AOE-4) Med.-Gulf September 2001–March 2002

HC-8 Det. 5 USS *Wasp* (LHD-1) February–August 2004

HC-8 Det. 7 USS *Arctic* (AOE-8) Med. & Gulf 1999

HC-8 Det. 5 USS *Concord* Summer Med. 1986

HC-8 Team 6 1994

HC-8 Det. 7 USS *Kearsarge* (LHD-3) ASAR, Med. 2001

HC-8 Det. 7 Raiders

HC-9 Strike Rescue

HC-11 Gunbearers (Desert)

HC-9 Search and Rescue

HC-9 Combat Rescue

HC-11 Flight Deck USS *Sacramento* (AOE-1) Westpac 1997–1998

HC-9 Combat Search and Rescue

HC-11 Gunbearers

HC-11 Junior Officers Protective Association

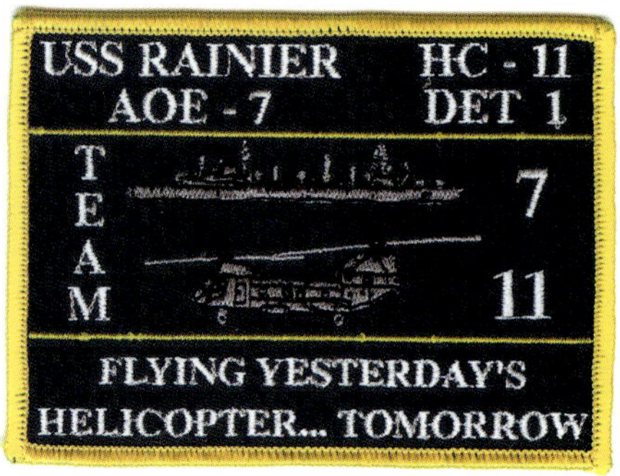

HC-11 Det. 1 USS *Rainier* (AOE-7), Team 7 & 11

HC-11 First SAR Det. 4 USS *Tarawa* (LHA-1) Westpac 1992

HC-11 Det. 5 Vacationing 1989

HC-11 Det. 2 USS *Camden* (AOE-2) Westpac, Operation Iraqi Freedom 2002–2003

HC-11 Det. 4 USS *Tripoli* (LPH-10) Worstpac June 9–December 9, 1994

HC-11 Det. 6, 1992

HC-11 Det. 3 Operation Desert Shield

HC-11 Det. 5 July–December 1976

HC-11 Det. 6 USS *Peleliu* (LHA-5) Operation Enduring Freedom

HC-11 Det. 7 Operation Desert Storm 1990–1991

HC-11 Det. 8 USS *Sacramento* (AOE-1) Operation Desert Storm

HC-11 Det. 8 Operation Desert Storm 1990–1991, Kuwait Liberation

HC-11 Det. 8 USS *Sacramento* (AOE-1)

HC-11 Det. 8 USS *Sacramento* (AOE-1) Westpac 1990–1991

HC-11 Det. 9 USS *Peleliu* (LHA-5) ESG-1 The Experiment 2003–2004

HC-11 Det. 8 Pacex September 14–December 16, 1989, Rimpac March 19–May 25, 1990, Northpac August 14–September 18, 1990, Westpac November 1990–December 1991

HC-11 Det. 8 Operation Desert Storm, Kuwait Liberation 1990–1991

HC-16 Search and Rescue

HC-16

HC-16 Det. 3 Operation Blue Harrier 1993

HC-85 (Desert)

HC-16

HC-28 USS *Nassau* (LHA-4) 2002–2003

HC-85 H-3 Sea King Plane Captain

HC-16

HC-85

HCS-4 Redwolves

ICS-4 Redwolves (Subdued)

ICS-4 Det. 1 Strike Rescue Operation Desert Shield

ICS-4 Operation Iraqi Freedom

HCS-4 Operation Iraqi Freedom

HCS-5 Firehawks

HCS-5 Firehawks (Subdued)

HCS-5 Firehawks (Desert)

HCS-5 Det. 1 Operation Iraqi Freedom, Phase III

HCT-16

HM-12

HM-12 Vertical Onboard Delivery Det.

HM-14 Horse Head

HM-12

HM-14

HM-14 Horse Head (Subdued)

HM-12

HM-14 (Subdued)

HM-14 Horse Head (Desert)

M-14 Horse Head (Cherry Blossoms in Japan)

HM-14 H-53E Plane Captain (Subdued)

HM-14 USS *Tripoli* (LPH-10) Westpac 1983

M-14 AMCM First Crewman

HM-14 MK-105 Sled Captain

HM-14 Persian Gulf 1987–1988

M-14 AMCM First Crewman (Subdued)

HM-14 FRS Instructor

HM-14 USS *Inchon* (LPH-12) Op. Blue Harrier 1993

HM-14 Endless Summer

HM-14 Southbound Trooper

HM-14 USS *Denver* (LPD-9) Rimpac 2006

HM-14 CJTFEX, Rimpack 2004

HM-14 The Great Himalayan Flying Club

HM-14 USS *Comstock* (LSD-45) Rimpac 2008

HM-14 USS *Bataan* (LHD-5) Panamax 2005

HM-14 USS *Kearsarge* (LHD-3) Panamax 2006

HM-14 Big Iron Productions

HM-14 USS *Boxer* (LHD-4) 2009

HM-14 Operation Trident Fury

HM-14 Det. 2 USS *Iwo Jima* (LHD-7) Hurricane Katrina

HM-14 Loss of Crew and Aircraft 543 January 08, 2014

HM-14 Det. 1 (Subdued)

HM-14 AMCM Unit Bravo

HM-14 Loss of Crew and Aircraft 543 January 08, 2014 (Subdued)

HM-14 Det. 2 Get the X

HM-15

HM-15

HM-15 (Subdued)

HSL-30 Kaman SH-2F Seasprite Buno. 159026, Modex 030.

HM-15 (Desert)

HM-15 AMCM

HM-15 Since 1987

HM-15 Plank Owner

HM-15 AMCM Tactical Support Division, RHTB Coxswain

HM-15 Blackhawks Motorcycle Riders Group

HM-15 Sled Captain

HM-15 AMCM 1st Crewman

HM-15 Pacex 1989

HM-15 Black Hawks

HM-15 Det. 2

HM-15 USS *Comstock* ((LSD-45) Rimpac 2010

HM-15 Black Hawks (Desert)

HM-15 Det. 2 (Desert)

HM-16

HM-15 Operation Freedom Banner 1995

HM-15 Det. 2 USS *Ponce* (AFSB-1) 2014

HM-16 Operation Blue Harrier, Denmark 1985

HM-18 Norsemen

HM-18 Norseman

HM-19 Golden Bears

HM-18 Norsemen

HM-18

HM-19 H-53 Plane Captain

HM-18

HM-18 Maintenanceman

HM Squadrons' 40th Anniversary

HS-1

HS-1 Seahorses

HS-1 Rappel

HS-1 Forty Year Anniversary 1951–1991

HS-1 Det. 1 Desert Duck Airlines

HS-1 Det. 3 Golden Chariots

HS-2

HS-2 (Subdued)

HS-2

HS-2 (Desert)

HC-2 Fifty Years 1952–2002

HS-2 Falcons

HS-2 Golden Falcons

HC-2 Fifty Year 1952–2002

HS-2- 100 Non-Arrested Carrier Landings

HS-2 Sikorsky Superior Helicopter Maintenance Award 1987

HS-2 SH-60, HH-60H Plane Captain

HS-2 USS *Abraham Lincoln* (CVN-72) Westpac 2006

HS-2 Det. A USS *Harry W. Hill* (DD-986) November 05-23 1987

HS-3 Tridents

HS-3 Battle Readiness "E" 1999

HS-3

HS-3 Combat SAR (Desert)

HS-3 Fifty Years 1952–2002

HS-3

HS-3 Combat SAR & Strike Seals

HS-3 Battle Readiness "E" 2008

HS-3 Trident Pirate Hunt 2009

HS-4 Black Knights

HS-4 Helicopters from Hell

HS-4

HS-4 Black Knights (Desert)

HS-4 Black Knights Sikorsky Excellence in Maintenance Award 1994

HS-4 Black Knights

HS-4 Unit Baker USS *Rendova* (CVE-114) Far East

HS-4 Combat Search and Rescue (Subdued)

HS-4 Navy Combat Search and Rescue (Subdued)

HS-4 Support Combat Search and Rescue (Desert)

HS-5 Nightdippers

HS-4 Millennium Knight 2000

HS-4 USS *Ronald Reagan* CVN-76) Basrah, Iraq 2006 (Desert)

HS-5 Nightdippers

HS-4 USS *Abraham Lincoln* (CVN-72) Westpac Operation Enduring Freedom and Southern Watch 2002

HS-5 Nightdippers

HS-5 Nightdippers

HS-5 Nightdippers

HS-5 Nightdippers (Desert)

HS-5 Med. Cruise 1984–1985

HS-5 Nightdippers

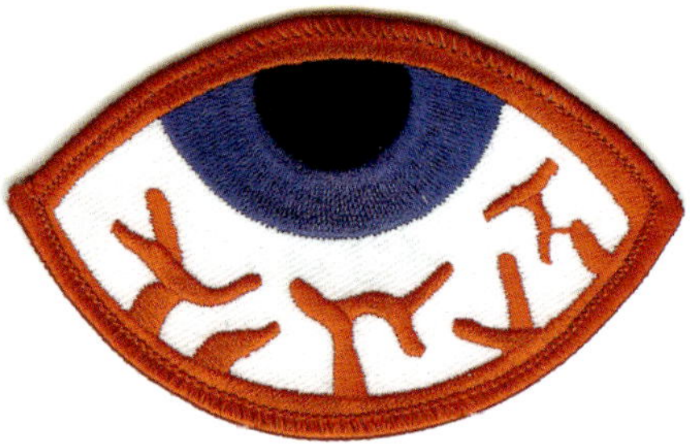
HS-5 Nightdippers Eye

HS-5 Battle Efficiency E 1992–1993

HS-5 Nightdippers (Subdued)

HS-5 Small Deck Centurion

HC-5 Battle Efficiency E 1996

HS-5 Federally Subsidized Helicopter Gang, Operation Enduring Freedom

HS-6

HS-6

HS-5, 50th Anniversary 1956–2006

HS-6

HS-6

HS-6

HS-6

HS-6 (Subdued)

HS-6 (Desert)

HS-6 Indian

HS-6 Westpac 2005 Operation Iraqi Freedom

HS-6 Indian (Subdued)

HS-6 Indian (Desert)

HS-6 Operation Enduring Freedom & Operation Iraqi Freedom, Westpac 2007

HS-6 Indian

HS-6 Tiger Cruise 1984

HS-6 Loss of Crew and Helicopter 617 May 19, 2009

S-7

S-7

S-7

HS-7 SH-3 Sea King

HS-7 SH-3H Sea King

HS-7 Operation Desert Storm 1990–1991

HS-7 Night dipping 100 Centurion

HS-7 Battle Readiness "E" 1994

HS-7 Med.-Persian Gulf, SH-3H Sunset Cruise 1994–1995

HS-7 Old No. 7 Brand

HS-7 Dusty Dogs

HS-7 Dusty Dogs (Desert)

HS-7 Forty Year Anniversary 1956–1996

HS-7 Dusty Dog Plane Captain

HS-7 USS *John C. Stennis* (CVN-75) Operatio
Iraqi Freedom & Med. 2003

HS-7 Around the World Club

HS-7 Dusty Dogs

HS-7 SH-60F & HH-60H

HS-8

HS-8

VC-5 Sikorsky SH-3A Buno. 148050, Tail Code UE, Modex 40, recovering a drone, January 10, 1986.

HS-8 (Subdued)

HS-8 (Desert)

HS-8 Eight Ball

HS-8 H-60 in front of Eight Ball

HS-8 SAR Det. 4 USS *Kitty Hawk* (CVA-63) March 1967

HS-8 Apollo Cruise October–November 1967

HS-8 Maintenance USS *Kitty Hawk* (CV-63) Westpac 1977–1978

HS-8 Sikorsky Excellence in Maintenance Award

HS-8 Det. Camp Arifjan, Operation Iraqi Freedom 2005

HS-8 World Cruise, Golden Wrench Award & Battle Readiness "E" Award

HS-9 Sea Griffins

HS-9 USS *Nimitz* (CVN-68) Med. Cruise 1987, End of an Era

HS-9 Sea Griffins

HS-9 Royal Order of the Palm, NATO 1966

HS-9 Med. 1989

HS-9 Sea Griffins

HS-9 Det. 66 USS *America* (CVA-66) Med. 1967

HS-10

HS-10

HS-10 Sikorsky SH-2 Excellence in Maintenance Award

HS-11

HS-10 Warhawks

HS-10 Sikorsky SH-3 Excellence in Maintenance Award

HS-11

HS-10 Warhawks (Subdued)

HS-10 SH-60F Warhawk Maintenance

HS-11 Dragonslayers (Subdued)

HS-11 Dragonslayers (Desert)

HS-11 SH-60 Plane Captain & LSE

HS-11 Battle Readiness "E" 1997

HS-11 Dragonslayers Plane Captain

HS-11 SH-60 Plane Captain

HS-11 Battle Readiness "E" 2006

HS-11 Battle Readiness "E" 1993

HS-11 Battle Readiness "E" 2007

HS-11 Joe Camel

HS-11 Det. 10 USS *Ogden* LPD-5 2003–2004

HS-12 USS *Independence* (CV-62) Westpac 1962

HS-11 Black Flag Det. Basrah, Iraq 2006

HS-12

HS-12 Det. B Team Oly USS *Oldendorf* (DD-972) 1986

HS-11 Pirate Slayers

HS-12

HS-12 Top Torpedo 1987

HS-12 Top Torpedo 1988 | **HS-14** | **HS-14 Chargers (Subdued)**

HS-12 Strike Rescue | **HS-14 Chargers** | **HS-14 Plane Captain**

HS-13 Sub-Choppers | **HS-14 Chargers** | **HS-14 Goat Locker**

HS-14 World Famous Chargers

HS-14 Operation Southern Watch 1992 (Desert)

HS-14 Det. A That's How We Roll

HS-14 Operation Desert Storm

HS-14 USS *Kitty Hawk* (CV-63) 2003

HS-14 Det. Alpha, *Chosin* Ones

HS-14 Operation Desert Storm

HS-14 Striking Ladies

HS-14 Det. Alpha, *Chosin* Ones (Subdued)

HS-14 Det. B USS *Lassen* (DDG-82)

HS-15

HS-15 Battle Readiness "E" 1990

HS-14 Det. 5 Operation Iraqi Freedom

HS-15 (Subdued)

HS-17

HS-15 SH-60F Sea Hawk, Dip Hard, Fast & Often

HS-15 (Desert)

HS-17 Neptune's Raiders Med. Cruise 1985-1986

HS-74

HS-75 Emerald Knights

HS-75 USS *Constellation* (CV-64), Shellback

HS-75 Emerald Knights

HS-75 Emerald Knights

HS-80L2 NAS Los Alamitos, CA

HS-75 Emerald Knights

HS-75 Emerald Knights

HS-84

HS-85

HS-246 Pterodactyls

HS-752 NARTU, NAS Lakehurst, NJ

HS-85

HS-742 NARTU, NAS Jacksonville, FL

HS-773 NAS Los Alamitos, CA

HS-85 Plane Captain

HS-751 NARTU, NAS Lakehurst, NJ

HS-831 NAS New York, NY

HS-892 NAS Seattle, WA

HSC-2 MH-60S FRS Instructor

HSC-3 (Desert)

HSC-2 Fleet Angels

HSC-2 MH-60S Plane Captain

HSC-3 40th Anniversary,
237,000 Flight Hours

HSC-2 Fleet Angels (Subdued)

HSC-3

HSC-4 Black Knights

HSC-6

HSC-6 (Desert)

HSC-7 Plank Owner

HSC-6 (Subdued)

HSC-6 Screamin' Indians

HSC-7 Plank Owner, *Saipan* March 2013

HSC-6 (Subdued)

HSC-6 Loss of Crew and Helicopter 617 September 22, 2013

HSC-7 In Memory of Lost Crew & Helicopter 612, May 7, 2007

HSC-7

HSC-7 Puppy

HSC-8 Eightballers

HSC-7 (Subdued)

HSC-7

HSC-8 Eightballers (Pink)

HSC-7 (Desert)

HSC-7 Dusty Dogs, MH-60S Armed Helicopter

HSC-8 Eightballers (Subdued)

HSC-8 Eightballers (Desert)

HSC-8 Det. 1 All Stars CG-53

HSC-8 Det. 1

HSC-8 Eightballers Mace

HSC-8 Det. 1 Jesters CG-53 2011-2012

HSC-8 Det. 2 USNS *Mercy* (T-AH-19)

HSC-8 Operation Podium & Operation Greenflash 2010

HSC-8 Det. 1 USNS *Bridge* (T-AOE-10)

HSC-9

HSC-9 (Desert)

HSC-9 Special Operations

HSC-12 Golden Falcons

HSC-9 Search & Rescue

HSC-9 Fallon Weapons Det.

HSC-12

HSC-9 Fortus ut Unim United Fortior 1913

HSC-9 Det. 1 USNS *Sacagawea* (T-AKE-2) 2011

HSC-12 Plank Owner

HC-5 Providers Boeing CH-46A Sea Knight conducting emergency personnel transfer hoist training from a Los Angeles class submarine in Apra Harbor, Guam.

HSC-12 MH-60S

HSC-12 CSAR, Elvis

HSC-21 Blackjacks

HSC-21 Blackjacks (Subdued)

HSC-21 The A Team Naval Air Ambulance Det.

HSC-21 Det. 3 USS *Peleliu* (LHA-5) Westpac 2006

HSC-21 Blackjacks (Desert)

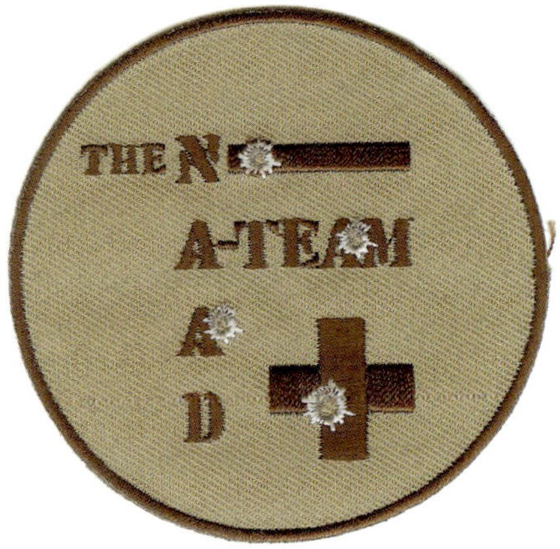

HSC-21 The A Team Naval Air Ambulance Det. (Desert)

HSC-21 Det. 3 Surrender the Booty

HSC-21 Blackjacks (Desert)

HSC-21 Det. 2 USNS *Mercy* (T-AH-19) Pacific Partnership. Conducted Humanitarian Missions to Indonesia, Philippines, Vietnam, and Cambodia 2012

HSC-21 Det. 3 USS *Boxer* (LHD-4) Amphibious Readiness Group 2013–2014

HSC-21 Det. 4 Fujairah to Mairah

HSC-22 Sea Knights (Subdued)

HSC-22 Rhino Empress

HSC-21 Det. 5 Air Ambulance (Desert)

HSC-22 Sea Knights (Desert)

HSC-22 Det. 1 Starboard Defenders

HSC-22 Sea Knights

HSC-22 Sea Knights Plankowner
October 1, 2006

HSC-22 Det. 1 Going Nowhere Fast

HSC-22 Det. 1 Deuces Wild

HSC-22 USNS *Arctic* (T-AOE-8) Polar Knights

HSC-22 SAR Det.

HSC-22 Det. 1

HSC-22 Det. 3 USNS *Supply* (T-AOE-6)

HSC-23 Wildcards

HSC-22 Det. 2 USS *Freedom* (LCS-1) Keep Back 8,000 M

HSC-22 Det. 3 Bird Dogs

HSC-23 Wildcards (Subdued)

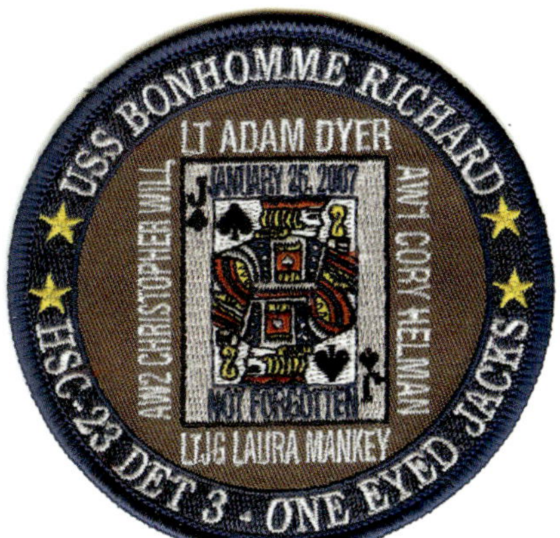

HSC-23 Wildcards (Desert)

HSC-23 Arrogant Vertrep

HSC-23 Det. 2 USNS *Bridge* (T-AOE-10) (Desert)

HSC-23 SH-60S The Sierra Club

HSC-23 Det. 1 2515th Naval Air Ambulance Det.

HSC-23 Det. 3 USS *Bonhomme Richard* (LHD-6)

HSC-23 Crossed Canes

HSC-23 Det. 2 USNS *Bridge* (T-AOE-10)

HSC-23 Det. 3 One Eyed Jacks Loss of Crew USS *Bonhomme Richard* (LHD-6) January 26, 2007

HSC-23 Det. 3 USS *Bonhomme Richard* (LHD-6)

HSC-23 Det. 4 Purple Pandas (Subdued)

HSC-23 Combat SAR Det. 5

HSC-23 Det. 3 USS *McClusky* (FFG-41) CTOC One Panga at a Time

HSC-23 Det. 4 Purple Pandas (Desert)

HSC-23 SAR Det. 5 Armed and Dangerous

HSC-23 Det. 4

HSC-23 SAR Det. 5 USS *Peleliu* (LHA-5) 2010

HSC-25 Island Knights

HSC-25 Island Knights (Subdued)

HSC-25 Island Knights (Desert)

HSC-25 Search and Rescue

HSC-25 Search and Rescue (Subdued)

HSC-25 Island Knight Pride

HSC-25 Det. 1 USS *George Washington* (CVN-73) USS *Flint* (T-AE-32) and USS *Shasta* (T-AE-33) November 2008–November 2009

HSC-25 Det. 1 USS *Essex* (LHD-2) April–December 2007

HSC-25 Det. 1 USS *George Washington* (CVN-73) May–December 2010

HSC-25 Det. 1 USNS *Charles Drew* (T-AKE-10) February– December 2013

HSC-25 Det. 4 USNS *Kiska* (T-AE-35) and USNS *Flint* (T-AE-32)

HSC-25 Det. 6 Porky's Revenge 2009

HSC-25 Det. 1 USNS *Charles Drew* (T-AKE-10) February–December 2013

HSC-25 Det. 5 Bullet Air

HSC-25 Det. 6 Top Hats & Mustaches

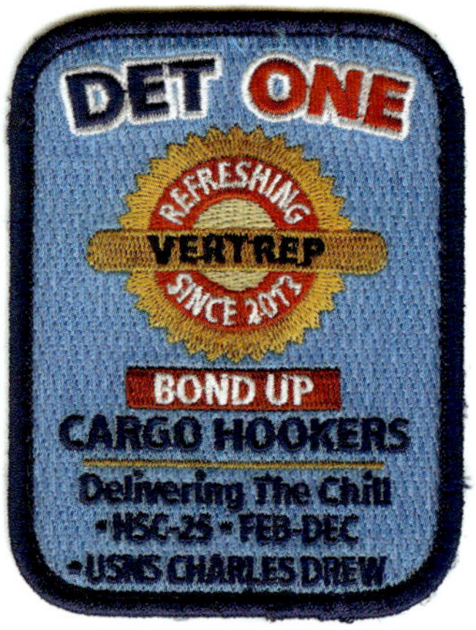

HSC-25 Det. 1 USNS *Charles Drew* (T-AKE-10) February–December 2013

HSC-25 Det. 6 Iron Gator SAR 2008

HSC-25 Det. 6 Top Hats & Mustaches (Subdued)

HSC-25 Det. 6 Armed Search and Rescue

HSC-25 Island Knights Always on Watch

HSC-26 (Subdued)

HSC-25 Det. 6 Strap on Another Package

HSC-26

HSC-26 (Desert)

HSC-25 Island Knights 30 Years. 1984–2014

HSC-26

HSC-26 Forty Years of Excellence September 1967–September 2007

HSC-26 Air Ambulance (Desert)

HSC-26 Medevac (Desert)

HSC-26 Junior Officers Protection Associatio

HSC-26 Medevac (Desert)

HSC-26 NATO Medevac

HSC-26 World Famous Desert Hawks (Desert)

HSC-26 Medevac (Desert)

HSC-26 Operation Enduring Freedom 2008–2009

HSC-26 Bill Murray Det.

HSC-26 Det. 1 Desert Hawks

HCS-26 Det. 1 Desert Hawks 2005 (Desert)

HSC-26 Det. 1 Hawk Air (Desert)

HSC-26 Det. 1 Desert Hawks (Desert)

HSC-26 Det. 1 Running the Gulf Since 2005 (Desert)

HSC-26 Det. 1 Desert Hawks (Desert)

HSC-26 Det. 1 Mail Escorts

HSC-26 Det. 1 Mafia (Desert)

HSC-26 Det. 1 USS *Bataan* (LHD-5) Search and Rescue 2014 (Desert)

HSC-26 Det. 2 USS *Saipan* (LHA-2) Euro trip 2005 (Desert)

HSC-26 Det. 2 USNS *Comfort* (T-AH-10)

HSC-26 Det. 2 USNS *Medgar Evers* (T-AKE-13)

HSC-26 Det. 2 USS *Saipan* (LHA-2) 2005 (Desert)

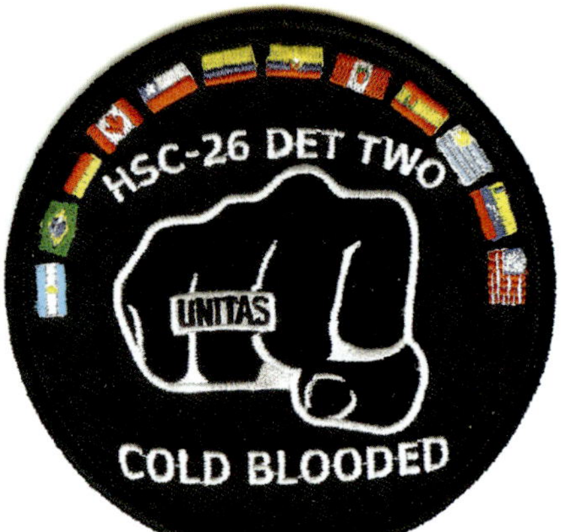

HSC-26 Det. 2 UNITAS

HSC-26 Det. 2 Laserhawks

HSC-26 Det. 2 Dealers

HSC-26 Det. 2 USNS *Medgar Evers* (T-AKE-13)

HSC-26 Det. 2 Laserhawks (Desert)

HSC-26 Det. 3 USNS *Comfort* (T-AH-10)

HSC-26 Det. 3 USNS *Comfort* (T-AH-10)

HSC-26 Det. 3 USS *Mesa Verde* (LPD-13) Pirates of the Med.

HSC-26 Det. 4

HSC-26 Det. 4 (Desert)

HSC-26 Det. 4 Get Lucky

HSC-26 Det. 4 Get Lucky (Desert)

HSC-26 Det. 4 USNS *Supply* (T-AOE-6) Cargo Ninjas

HSC-26 Det. 4 USNS *Supply* (T-AOE-6) Slingin' Meat To the Fleet

HSC-26 Det. 4 USNS *Supply* (T-AOE-6) Slingin the Bull

HSC-26 Det. 4 Gunrunners, Just Mark it P.I.

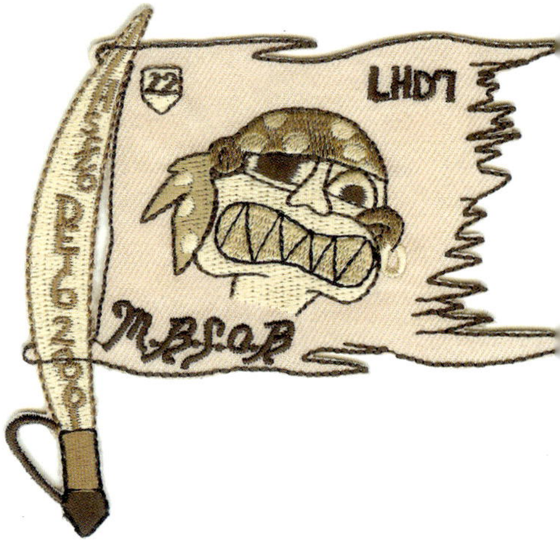

HSC-26 USS *Iwo Jima* (LHD-7) Flag (Desert)

HSC-26 Det. 4 Hurricane Sandy

HSC-26 Det. 4 USS *Wasp* (LHD-1)(F-35)

HSC-26 Det. 4 Medical

HSC-26 Det. 4 USS *Wasp* (LPD-1) NAS Patuxent River JSF, F-35

HSC-26 Det. 6 Choke on this

HSC-26 Det. 11 Airwolves

SC-28

HSC-28 (Desert)

HSC-28 Operation Warrior Kraken March 2011–February 2012

SC-28 (*Gray*)

HSC-28 Combat Readiness "E" Award 2011

HSC-28 It Is What It Is

SC-28 (Subdued)

HSC-28 Plane Captain

HSC-28 Endless Summer Cruise 2013

HSC-28 Combat Readiness "E" Award 2013

HSC-28 Det. 1 Ghostriders

HSC-28 Det. 1 and Det. 2 (Rein) 2012

HSC-28 Twenty-Five Year Anniversary 1984–2009

HSC-28 Det. 1 Ghostriders Napoli, Italy

HSC-28 Det. 1 and Det. 2 (Rein) (Subdued) 201

HSC-28 Det. 1 Ghostriders

HSC-28 Det. 1 Ghostriders

HSC-28 Det. 2 March 23, 2011–?

HSC-28 Det. 2 USNS *Comfort* (T-AH-20) Haitian Vacation

HSC-28 Det. 4 USNS *Arctic* (T-AOE-8)

HSC-28 Det. 5 / MSRT Det. 1 VooDoo Flyers May 2014

HSC-28 Det. 3 The Savages

HSC-28 Det. 4 USS *Nassau* (LHA-4) 2009–2010

HSC-28 Det. 5 / MSRT Det. 1 VooDoo Flyers May 2014 (Subdued)

HSC-28 Det. 3 USS *Nassau* (LHA-4) 2008

HSC-28 Det. 5 VooDoo Knights

HSC-28 Det. 8 On the Ball

HSC-28 Det. 8

HSC-84 Red Wolves (Subdued)

HSC-85 Firehawks

HSC-37 Det. 3 USS *Lake Erie* (CG-70)
Rimpac 2008

HSC-84 Red Wolves (Desert)

HSC-85 Firehawks (Subdued)

HSC-84 Red Wolves

HSC-84 Red Wolves (Desert)

HSC-85 Firehawks (Desert)

HSC-85 High Rollers

Multi-Squadron; HSC-9, HSC-22, HSC-26 & HSC-28 Haiti the Game 2010

HSC Blue Cherubs, Squadron Unknown

HSC-85 High Rollers

Multi-Squadron; HSC-2 & HSC-9 USS *San Antonio* (LPD-17) Hurricane Sandy 2012

HSL-30

HSC-85 High Rollers Combat Readiness "E" Award 2007

Multi-Squadron; HSC-25 & HS-15 Naval Air Ambulance Det.

HSL-30

HSL-30 Det. Alfa 15

HSL-31 Det. Bravo Panama 1977–1978

HSL-32 Det. 1 Mooncrickets USS *John Rodgers* (DD-983)

HSL-31 Arch Angels

HSL-31 Det. Bravo Arch Angels

HSL-32 Det. 2 USS *Capodanno* (FF-1093) Med. Cruise January 1990

HSL-31 Det. A Fleet Air Rapid Transit

HSL-32

HSL-32 Det. 3 USS *O'Bannon* (DD-987) 1984

HSL-32 Det. 3 USS *Ticonderoga* (CG-47) Med. February 1990

HSL-32 Det. 6 USS *Comte De Grasse* (DD-974) North Atlantic-Med. Cruise 1981–1982

HSL-32 Det. 8 USS *Truett* (FF-1095)

HSL-32 Det. 5 USS *Capodanno* (FF-1093) Counter Drug Operations 1992

HSL-32 Det. 6 USS *Josephus Daniels* (CG-27)

HSL-32 Det. 8 Anytime Buckwheat

HSL-32 Det. 5 USS *Capodanno* (FF-1093) Med. Cruise January 1993

HSL-32 Det. 8 Maintenance Crew

HSL-32 Det. Zulu Ground Pounders

HSL-33

HSL-33 Det. 5 Hell's Belles

HSL-33 Det. 7 Persian Gulf 1990

HSL-33 SH-2F Bad to the Boneyard

HSL-33 Det. 5 Drug Bustin 1991–1992

HSL-33 Det. 10 A Perfect 10

HSL-33 Det. 2 Operation Praying Mantis April 18, 1988

HSL-33 Det. 7 Rescue

HSL-34

HSL-34

HSL-34 Year of Excellence Safety "S" Award and Battle Readiness "E" Award 1992

HSL-34 Det. 1 Leaders of the Pack

HSL-34 USS *Caron* (DD-970) Med. Cruise, Black Sea Operations March 1985

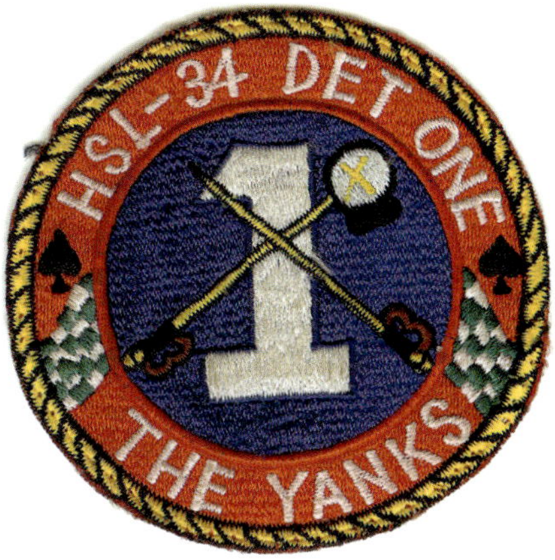

HSL-34 Det. 1 The Yanks

HSL-34 Det. 1 USS *Biddle* (CG-34) Operation Desert Storm August 1990–March 1991

HSL-34 SH-2F Plane Captain

HSL-34 Det. 1 Traveling Maintenance Show

HSL-34 Det. 2 USS *Hayler* (DD-997) UNITAS XXVII 1986

HSL-34 Det. 5 Seasnakes

HSL-34 Det. 8 Crusaders

HSL-34 Det. 6 UNITAS XXXII

HSL-34 Det. 10 USS *Scott* (DDG-995) Sled Dogs
Med. February 1990

HSL-34 Det. 8 The Real Thing 1983–1984

HSL-34 Det. Whiskey

HC-11 Gun Bearers Boeing CH-46A Sea Knight Buno. 150951, Tail code VR, Modex 46, 100,000 Mishap Free Flight Hours.

HSL-35

HSL-35 Det. 5 USS *Gray* (FF-1054) Westpac 1977–1978

HSL-35 Det. 7 Operation Desert Storm August 2–November 2, 1990

HSL-35

HSL-35 Det. 5 Team A Westpac 1983–1984

HSL-36

HSL-35 Det. 1 Pirates

HSL-35 Det. 6 Lamps USS *Kinkaid* (DD-965)

HSL-36 Awards Clean Sweep 1990

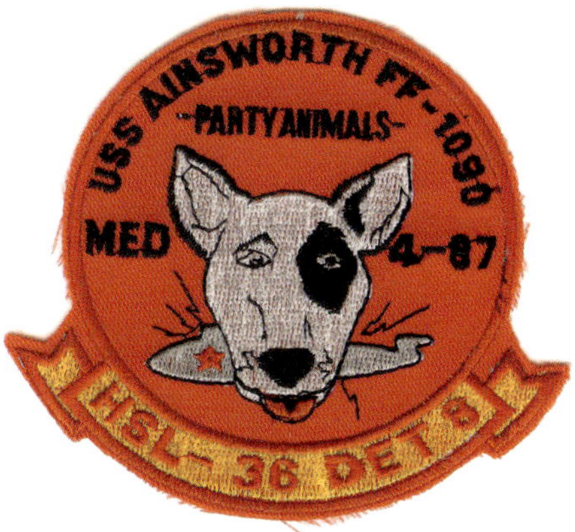

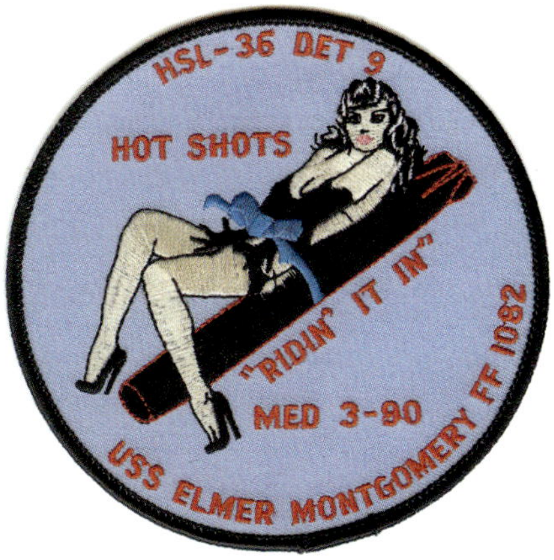

HSL-36 Det. 2 USS *Nicholson* (DD-982) Rhumb Runners

HSL-36 Det. 3 USS *Bowen* (FF-1079) Med. March 1987

HSL-36 Det. 4 USS *Gallery* (FFG-26) Middle East Force February 1989

HSL-36 Det. 6 USS *Stark* (FFG-31) SNFL January-July 1990

HSL-36 Det. 7 USS *Paul* (FF-1080) Med. February 1990

HSL-36 Det. 8 USS *Ainsworth* (FF-1090) Med. April 1987

HSL-36 Det. 8

HSL-36 Det. 9 Med. Cruise February–August 1988

HSL-36 Det. 9 USS *Elmer Montgomery* (FF-1082) Med. March 1990

HSL-37

HSL-37 Easy Riders (Desert)

HSL-37 Det. 1 Westpac 1984

HSL-37

HSL-37 Easy Riders Plane Captain

HSL-37 Operation Desert Storm, Iraqpac 1991

HSL-37

HSL-37-25 Years

HSL-37 Det. 1 Easy Riders USS *Port Royal* (CG-73)

HSL-37 Det. 1 USS *Port Royal* (CG-73) Westpac 2000

HSL-37 Det. 1 Squad "A"

HSL-37 Det. 1 Liquid Aloha (Desert)

HSL-37 Det. 1 Ichiban Express

HSL-37 Det. 1 Liquid Aloha

HSL-37 Det. 1

HSL-37 Det. 1 USS *Chafee* (DDG-90) Operation Foal Eagle and Operation Northern Edge 2008

HSL-37 Det. 1 Liquid Aloha (Subdued)

HSL-37 Det. 1 USS *Chafee* (DDG-90) Westpac 2010

HSL-37 Det. 1 USS *Chafee* (DDG-90) Westpac 2010 (Subdued)

HSL-37 Det. 1 USS *Chafee* (DDG-90) Westpac 2012 (Subdued)

HSL-37 Det. 1 USS *Chafee* (DDG-90) Westpac 2012 (Desert)

HSL-37 Det. 1 Defenders of Aloha

HSL-37 Det. 1 USS *Chosin* (CG-65) Westpac 2013

HSL-37 Det. 1 USS *Chosin* (CG-65) Westpac 2013 (Subdued)

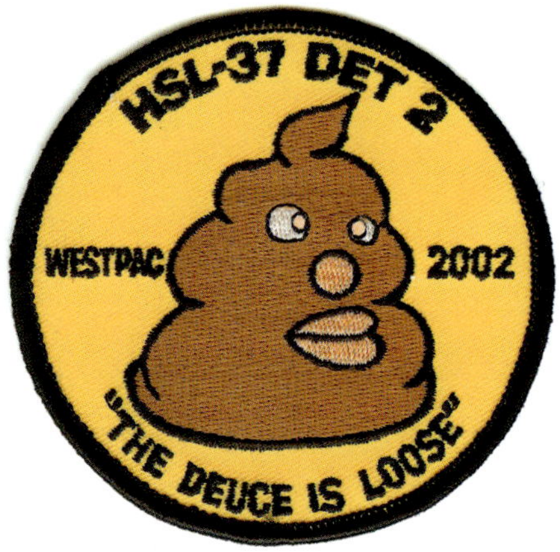

HSL-37 Det. 2 Westpac 2002

HSL-37 Det. 2 Westpac 2002 (Desert)

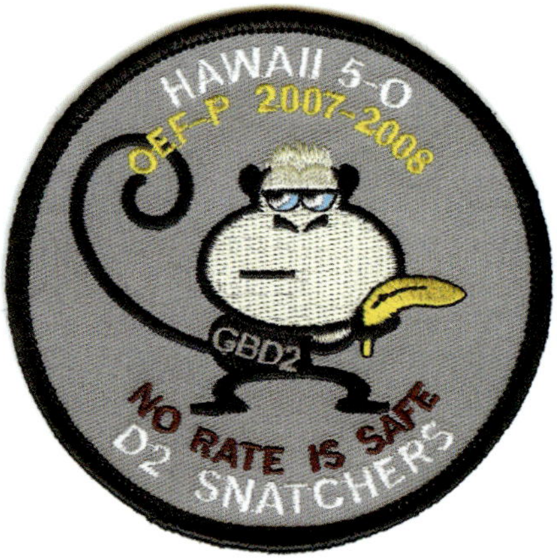

HSL-37 Det. 2 Operation Enduring Freedom 2007–2008

HC-8 Dragon Whales Boeing UH-46D Sea Knight Buno. 153405, Tail code BR, USS *Mt. Baker* (AE-34).

HSL-37 Det. 2 USS *Reuben James* (FFG-57)

HSL-37 Det. 2 USS *Ingraham* (FFG-61)
2009–2010

HSL-37 Det. 2 USS *Ingraham* (FFG-61) 2009–2010 (Subdued)

HSL-37 Det. 2 USS *Ingraham* (FFG-61) 2009–2010 (Desert)

HSL-37 Det. 2 USS *Crommelin* (FFG-37)

HSL-37 Det. 2 USS *Sampson* (DDG-10) 2012

HSL-37 Det. 2 USS *Sampson* (DDG-10) 2012 (Subdued)

HSL-37 Det. 2 USS *Sampson* (DDG-10) 2012 (Desert)

HSL-37 Det. 3 USS *Truxton* (CGN-35)

HSL-37 Det. 3 The Smoking Holes

HSL-37 Det. 3 USS *Crommelin* (FFG-37) Westpac 1996

HSL-37 Det. 3 USS *Crommelin* (FFG-37) Middle East Force 1998

HSL-37 Det. 3 Jesters

HSL-37 Det. 3 USS *Chung Hoon* (DDG-93)

HSL-37 Det. 3 Persian Gulf 2000

HSL-37 Det. 3 Murphy's Law

HSL-37 Det. 3 USS *Reuben James* (FFG-57) 2010 (Subdued)

HSL-37 Det. 3 Pirates of the Caribbean, In Theater Summer 2004

HSL-37 Det. 3 Rimpac 2008

HSL-37 Det. 3 USS *Reuben James* (FFG-57) 2010 (Desert)

HSL-37 Det. 3 USS *Reuben James* (FFG-57)

HSL-37 Det. 3 USS *Vandegrift* (FFG-48)
Westpac 2012 (Desert)

HSL-37 Det. 4 USS *Reuben James*
(FFG-57) 1999

HSL-37 Det. 3 USS *Reuben James* (FFG-57)

HSL-37 Det. 4 USS *Crommelin* (FFG-37)

HSL-37 Det. 4 USS *Reuben James* (FFG-57)
Westpac 2002

HSL-37 Det. 3 USS *Vandegrift* (FFG-48)
Westpac 2012

HSL-37 Det. 4 USS *Reuben James* (FFG-57)
Carat 2000

HSL-37 Det. 4 USS *Chosin* (CG-65) 2005

HSL-37 Det. 4 USS *Gridley* (DDG-101)

HSL-37 Det. 4 The Entourage

HSL-37 Det. 4 The Entourage (Subdued)

HSL-37 Det. 4 USS *Chung Hoon* (DDG-93) Westpac 2011

HSL-37 Det. 4 USS *Chung Hoon* (DDG-93) Westpac 2011 (Subdued)

HSL-37 Det. 4 USS *Chung Hoon* (DDG-93) Westpac 2011 (Desert)

HSL-37 Det. 4 USS *Reuben James* (FFG-57) 2012-2013

HSL-37 Det. 4 USS *Reuben James* (FFG-57) 2012-2013 (Subdued)

HSL-37 Det. 4 Uncle Ty's Toilet Hooch

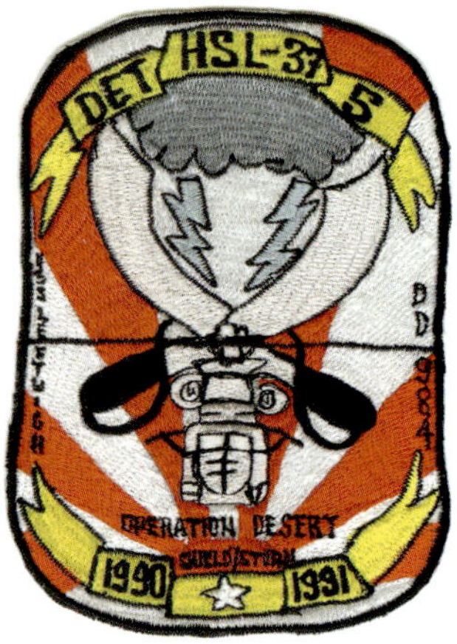

HSL-37 Det. 5 Operation Desert Shield & Operation Desert Storm 1990–1991

HSL-37 Det. 5 Earlybirdz

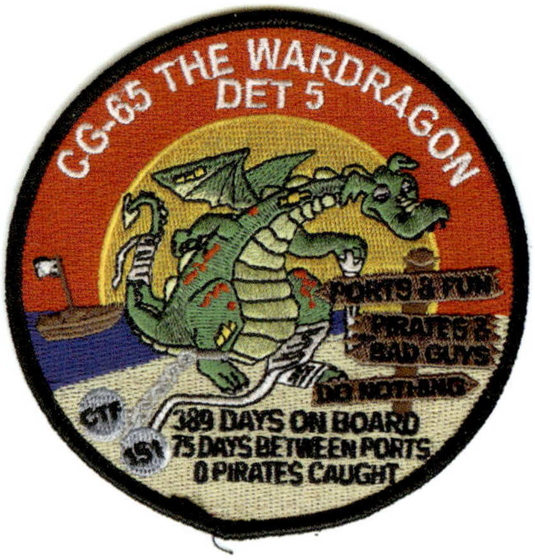

HSL-37 Det. 5 The Wardragon

HSL-37 Det. 5 Spies in the Sky

HSL-37 Det. 5 Earlybirdz (Subdued)

HSL-37 Det. 5 USS *Port Royal* (CG-73) Westpac & Arabian Gulf 2011-2012

HSL-37 Det. 5 USS *Ford* (FFG-54) Carat Cruise 2008 (Subdued)

HSL-37 Det. 5 Earlybirdz (Desert)

HSL-37 Det. 5 USS *Port Royal* (CG-73) Westpac & Arabian Gulf 2011-2012 (Subdued)

HSL-37 Det. 5 USS *Port Royal* (CG-73) Westpac & Arabian Gulf 2011-2012 (Desert)

HSL-37 Det. 5 USS *Chung Hoon* (DDG-93) Westpac 2013

HSL-37 Det. 5 USS *Chung Hoon* (DDG-93) Westpac 2013 (Subdued)

HSL-37 Det. 6 USS *Chosin* (CG-65) Westpac 1995

HSL-37 Det. 6 Dragon Riders USS *Chosin* (CG-65) Westpac 2001

HSL-37 Det. 7 Atsugi, Japan

HSL-37 Det. 7 USS *Fletcher* (DD-992) Westpac 2000

HSL-37 Det. 7 USS *Fletcher* (DD-992) We Deal in Lead

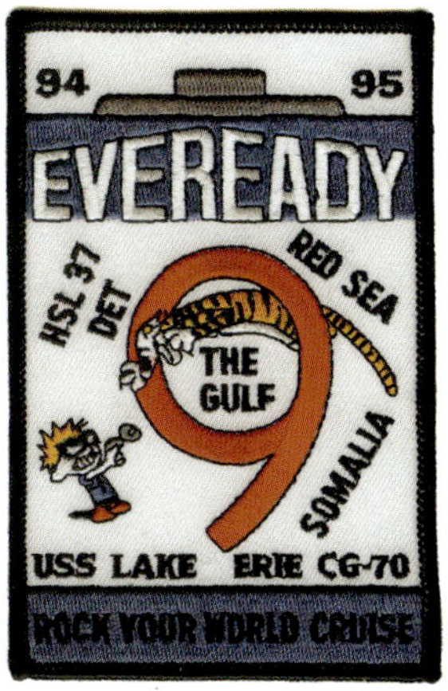

HSL-37 Det. 9 USS *Lake Erie* (CG-70) Westpac The Gulf, Red Sea, Somalia 1994-1995

HSL-37 Det. 9 USS *Lake Erie* (CG-70) Westpac 1997

HSL-37 Det. 10 USS *Cushing* (DD-905) Westpac 1982–1983

HSL-37 "U" Det. (Subdued)

HSL-37 Det. 9 USS *Lake Erie* (CG-70) Westpac 1999 (Subdued)

HSL-37 Det. Oscar

HSL-40 Airwolves

HSL-37 Det. 9 USS *Curts* (FFG-38) Carat 2001

HSL-37 "U" Det.

HSL-40 Airwolves

SL-40 Airwolves

HSL-40 Safety

HSL-41 Seahawks

SL-40 Airwolves

HSL-40 SH-60B Plane Captain

HSL-41 Seahawks

SL-40 Airwolves (Desert)

HSL-40 Ellyson Award Winners 1999–2001

HSL-41 Seahawks

HSL-41 Aircrew

HSL-42 Proud Warriors (Subdued)

HSL-42 SH-60B Plane Captain

HSL-42 Proud Warriors

HSL-42 Proud Warriors (Desert)

HSL-42 100 Hours Unaided (Night)

HSL-42 Proud Warriors

HSL-42 Work Center 010

HSL-42 USS *George Washington* (CVN-73) CQ Det. (Subdued)

SL-42 Safety "S" Award 1992–1994

HSL-42 Det. 1 USS *Underwood* (FFG-36) Med. Cruise 1985

HSL-42 Det. 1 Operation Desert Storm

SL-42 Combat Readiness "E" Award 1987, 1988, 1992, 1993, 2001

HSL-42 Det. 1 USS *Underwood* (FFG-36) Standard Naval Force Atlantic February 1987

HSL-42 Det. 1 USS *Caron* (DD-976) January 1996

SL-42 Combat Readiness "E" Award 1987, 1988, 1992, 1993, 2001

HSL-42 Det. 1 USS *San Jacinto* (CG-56) Med. & Black Sea March 1989

HSL-42 Det. 1 Fugawee Tribe USS *Monterey* (CG-61) Middle East Force 2002

HSL-42 Det. 1 USS *Normandy* (CG-60) Med. & Middle East Force 2005

HSL-42 Det. 1 USS *Vella Gulf* (CG-72) The Legends

HSL-42 Det. 1 Bad to the Bone

HSL-42 Det. 1 USS *Normandy* (CG-60) Med. & Middle East Force 2005 (Subdued)

HSL-42 Det. 1 USS *Vella Gulf* (CG-72) The Legends (Subdued)

HSL-42 Det. 1 USS *Nitze* (DDG-94) Operation Joint Warrior

HSL-42 Det. 1 USS *Hawes* (FFG-53) Med. Middle East Force 2000

HSL-42 Det. 1 USS *Vella Gulf* (CG-72) The Legends (Desert)

HSL-42 Det. 1 USS *Nitze* (DDG-94) Operation Joint Warrior (Subdued)

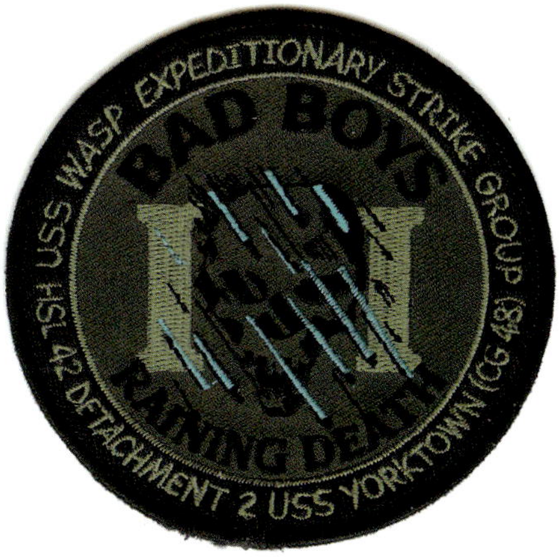

HSL-42 Det. 1 USS *Nitze* (DDG-94) Operation Joint Warrior (Desert)

HSL-42 Det. 2 USS *Elwood* (FFG-55) Med. Cruise 1987

HSL-42 Det. 2 Middle East Force March 1988

HSL-42 Det. 2 USS *Boone* (FFG-28) Counter Drug Operations & UNITAS

HSL-42 Det. 2 USS *Wasp* (LHD-1) & USS *Yorktown* (CG-48) Expeditionary Strike Group

HSL-42 Det. 2 USS *Wasp* (LHD-1) & USS *Yorktown* (CG-48) Expeditionary Strike Group (Subdued)

HSL-42 Det. 2 USS *Boone* (FFG-28) World Tour, Operation North Star, Med.- MIF- MEF 1991–1992

HSL-42 Det. 2 USS *Klakring* (FFG-42) Med. February 1996

HSL-42 Det. 2 USS *Nitze* (DDG-94) Jive Turkeys

USMC Camp Pendleton, CA, HSC-3 MH-60S specially equipped to conduct aerial water drops against wildfires in San Diego County, CA.

HSL-42 Det. 2 USS *Nitze* (DDG-94) Jive Turkeys (Subdued)

HSL-42 Det. 2 USS *Nitze* (DDG-94) Jive Turkeys (Desert)

HSL-42 Det. 2 USS *Spruance* (DD-953) Standard Naval Force Med. Into the Millennium 1999–2000

HSL-42 Det. 2 USS *Klakring* (FFG-42)

HSL-42 Det. 2 USS *Vicksburg* (CG-69) Surge 2006

HSL-42 Det. 2 USS *Vicksburg* (CG-69) Surge 2006 (Desert)

HSL-42 Det. 2 USS *Halyburton* (FFG-40) SNMG-1 2010

HSL-42 Det. 3 USS *Klakring* (FFG-42) Commander Middle East Force June–December 1985

HSL-42 Det. 3 USS *Leyte Gulf* (CG-55) Med. January 1989

HSL-42 Det. 3 USS *Leyte Gulf* (CG-55) Med. & Red Sea Cruise January 1991

HSL-42 Det. 3 USS *Leyte Gulf* (CG-55) Med. January 1993

HSL-42 Det. 3 USS *Leyte Gulf* (CG-55) Joint Task Group 1-97, 1997

HSL-42 Det. 3 USS *Moosbrugger* (DD-980) UNITAS 38-1998

HSL-42 Det. 3 USS *Ticonderoga* (CG-47) UNITAS 41- 2000

HSL-42 Det. 3 USS *John L. Hall* (FFG-32) Standard Naval Force Atlantic 2000–200

HSL-42 Det. 3 USS *Deyo* (DD-989) Med. Middle East Force 2003

HSL-42 Det. 3 USS *Kauffman* (FFG-59) 2005

HSL-42 Det. 3 USS *Roosevelt* (DDG-80) 2008–2009

HSL-42 Det. 3 USS *Roosevelt* (DDG-80) 2008–2009 (Subdued)

HSL-42 Det. 3 USS *Roosevelt* (DDG-80) 2008–2009 (Desert)

HSL-42 Det. 3 USS *Robert G. Bradley* (FFG-49) Counter Drug Operations & UNITAS

HSL-42 Det. 3 Naval Aviation Special Forces

HSL-42 Det. 3 USS *Hawes* (FFG-53) Fat Monkeys

HSL-42 Det. 3 USS *Hawes* (FFG-53) Fat Monkeys (Subdued)

HSL-42 Det. 3 USS *Vella Gulf* (CG-72) Three Norseman

HSL-42 Det. 4 USS *De Wert* (FFG-45) Med. March 1985

HSL-42 Det. 4 USS *Monterey* (CG-61) Millennium Cruise

HSL-42 Det. 3 USS *Vella Gulf* (CG-72) Three Norseman (Subdued)

HSL-42 Det. 4 USS *De Wert* (FFG-45) Middle East Force January 1989

HSL-42 Det. 4 USS *Yorktown* (CG-48) Counter Drug Operations & UNITAS, 2002

HSL-42 Det. 3 USS *Vella Gulf* (CG-72) Three Norseman (Desert)

HSL-42 Det. 4 USS *Samuel B. Roberts* (FFG-58) Middle East Force January 1998

HSL-42 Det. 4 USS *Spruance* (DD-963) Med. 2002

HSL-42 Det. 4 USS *James E. Williams* (DDG-95) Dirty Pirates

HSL-42 Det. 4 USS *James E. Williams* (DDG-95) Dirty Pirates (Subdued)

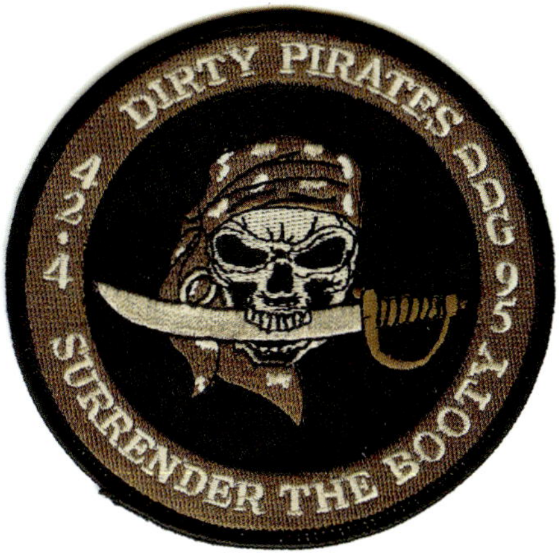

HSL-42 Det. 4 USS *James E. Williams* (DDG-95) Dirty Pirates (Desert)

HSL-42 Det. 4 Not Lucky, Just Good

HSL-42 Det. 4 USS *El Rod* (FFG-55) Lucky Devils

HSL-42 Det. 5 USS *Hawes* (FFG-53)

HSL-42 Det. 5 Operation Ocean Safari & Baltic Operations, Bluenose 1985

HSL-42 Det. 5 USS *Normandy* (CG-60) Operation North Star, Med. & Middle East Force 1991–1992

HSL-42 Det. 5 USS *Nicholas* (FFG-47) Dirty Dawgs, Standard Naval Force Atlantic February 1995

HSL-42 Det. 5 USS *Thorn* (DD-988) Night Vision Goggles Middle East Force February 1997

HSL-42 Det. 5 USS *Peterson* (DD-969) Standard Naval Force Med. 1999

HSL-42 Det. 5 USS *Nicholas* (FFG-47) Med. & Middle East Force January 2001 (desert)

HSL-42 Det. 5 USS *Anzio* (CG-58) Baltic Operations 2002

HSL-42 Det. 5 USS *Stump* (DD-978) Counter Drug Operations & UNITAS 2003

HSL-42 Det. 5 USS *Vicksburg* (CG-69) 2007

HSL-42 Det. 5 USS *Farragut* (DDG-99) USS *Eisenhower* Carrier Strike Group

HSL-42 Det. 5 USS *Farragut* (DDG-99) USS *Eisenhower* Carrier Strike Group (Subdued)

HSL-42 Det. 5 USS *Farragut* (DDG-99) USS *Eisenhower* Carrier Strike Group (Desert

HSL-42 Det. 5 Red Hand

HSL-42 Det. 6 USS *John L. Hall* (FFG-32) Night Vision Goggles, UNITAS 1996

HSL-42 Det. 7 USS *Hue City* (CG-66) Med. Cruise 1993

HSL-42 Det. 6 USS *Conolly* (DD-979) Pirates of the Caribbean

HSL-42 Det. 6 USS *Nicholas* (FFG-47) Summer 2005

HSL-42 Det. 7 USS *Kauffman* (FFG-59) Med. February 1995

HSL-42 Det. 6 USS *Moosbrugger* (DD-980) Med. January 1989

HSL-42 Det. 7 Warthogs

HSL-42 Det. 7 USS *Samuel B. Roberts* (FFG-58) Middle East Force & Med. 2000

HSL-42 Det. 7 USS *Vicksburg* (CG-69) Med. & Middle East Force 2001-2002

HSL-42 Det. 7 USS *John L. Hall* (FFG-32) Operation Neptune Warrior 2007

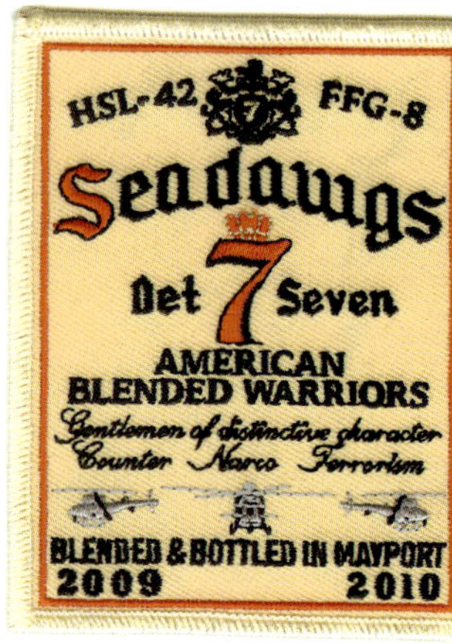

HSL-42 Det. 7 Seadawgs 2009–2010

HSL-42 Det. 7 USS *Gettysburg* (CG-64) USS *Enterprise* (CVN-65) Strike Group, 2003-2004

HSL-42 Det. 7 USS *Kauffman* (FFG-59) 0April–October 2008

HSL-42 Det. 8 USS *Vicksburg* (CG-69)

HSL-42 Det. 7 USS *Bulkeley* (DDG-84) Desert Cowboys 2006

HSL-42 Det. 7 USS *Kauffman* (FFG-59) April–October 2008 (Subdued)

HSL-42 Det. 8 USS *Nicholas* (FFG-47) Med. 1986

HSL-42 Det. 8 USS *McInerney* (FFG-8)

HSL-42 Det. 8 USS *San Jacinto* (CG-56) Med. & Middle East Force 2001

HSL-42 Det. 8 Mortal

HSL-42 Det. 8 USS *McInerney* (FFG-8) Med. 1993

HSL-42 Det. 8 USS *Hawes* (FFG-53) Med. 2002–2003

HSL-42 Det. 8 USS *Philippine Sea* (CG-58) What's Yours Is Mine (Subdued)

HSL-42 Det. 8 USS *De Wert* (FFG-45) Standard Naval Force Atlantic January 1999

HSL-42 Det. 8 USS *Carr* (FFG-52) Counter Narcotics Operations 2009

HSL-42 Det. 8 USS *Doyle* (FFG-39) Counter Narcotics Operations 2010-2011

HSL-42 Det. 8 USS *Doyle* (FFG-39) Counter Narcotics Operations 2010–2011 (Subdued)

HSL-42 Det. 8 USS *Doyle* (FFG-39) Counter Narcotics Operations 2010–2011 (Desert)

HSL-42 Det. 8 Damage Incorporated

HSL-42 Det. 8 Damage Incorporated (Subdued)

HSL-42 Det. 8 Damage Incorporated (Desert)

HSL-42 Det. 8 Possums

HSL-42 Det. 9 USS *Taylor* (FFG-50) Standard Naval Force Atlantic January–July 1987

HSL-42 Det. 9 USS *Thomas S. Gates* (CG-51) Escort Service

HSL-42 Det. 9 USS *Taylor* (FFG-50) Middle East Force March 1990

HSL-42 Det. 9 USS *Taylor* (FFG-50) Med. January 1994

HSL-42 Det. 9 USS *Nicholas* (FFG-47) Middle East Force 1997

HSL-42 Det. 9 USS *Hayler* (DD-997) UNITAS 41-2000

HSL-42 Det. 9 USS *Normandy* (CG-60) Joint Task Group March 1995

HSL-42 Det. 9 USS *Taylor* (FFG-50) Diamond Cutters

HSL-42 Det. 9 Counter Narcotics Operations & UNITAS 43-2002

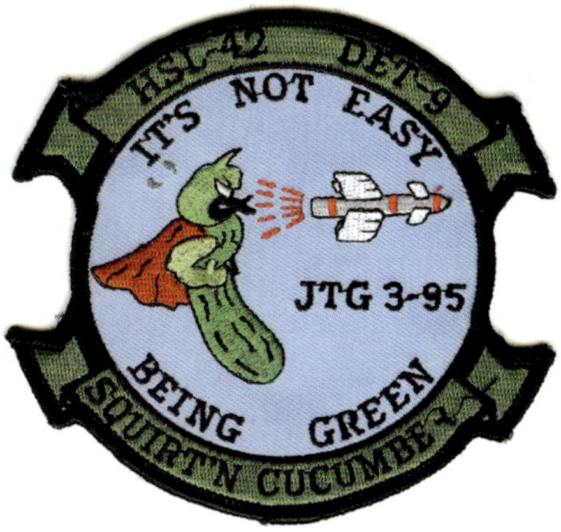

HSL-42 Det. 9 Joint Task Group March 1995

HSL-42 Det. 9 USS *Simpson* (FFG-56) LAMPS

HSL-42 Det. 9 USS *Gettysburg* (CG-64) Carrier Strike Group 2007

HSL-42 Det. 9 USS *Gettysburg* (CG-64) 9th Air Cavalry 2007

HSL-42 Det. 9 Flatliners USS *Thomas S. Gates* (CG-51)

HSL-42 Det. 10 Raging Roosters

HSL-42 Det. 9 USS *Gettysburg* (CG-64) 9th Air Cavalry 2007 (Subdued)

HSL-42 Det. 9 Ninjees USS *San Jacinto* (CG-56) 2010

HSL-42 Det. 10 USS *Oscar Austin* (DDG-79) ILL Tempered Sea Bass

HSL-42 Det. 9 USS *Gettysburg* (CG-64) 9th Air Cavalry 2007 (Desert)

HSL-42 Det. 9 USS *Nicholas* (FFG-47) Guns N Rotors

HSL-42 Det. 10 Hellraisers USS *Stephen W. Groves* (FFG-29)

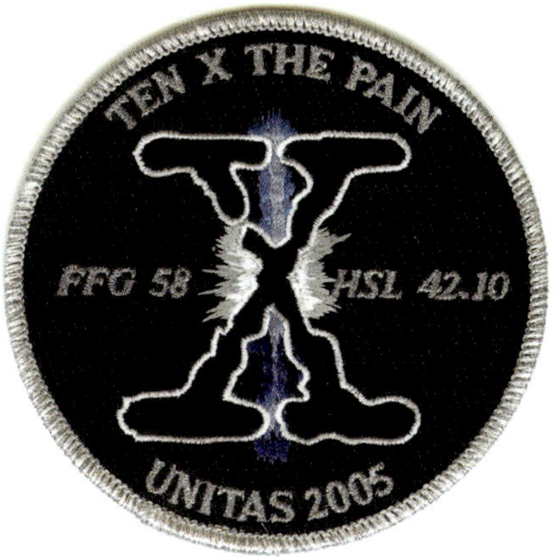

HSL-42 Det. 10 Med. January 1992

HSL-42 Det. 10 USS *Boone* (FFG-28) Med. 1995

HSL-42 Det. 10 USS *Carr* (FFG-52) Expeditionary Strike Group

HSL-42 Det. 10 Phist Pumpers USS *Klakring* (FFG-42) (Subdued)

HSL-42 Det. 10 Phist Pumpers USS *Klakring* (FFG-42) (Desert)

HSL-42 Det. 10 Maritime Interdiction Force January 1994

HSL-42 Det. 10 USS *James E. Williams* (DDG-95)

HSL-42 Det. 10 USS *Stephen W. Groves* (FFG-29) Counter Drug Operations & UNITAS 2003

HSL-42 Det. 10 USS *Samuel B. Roberts* (FFG-58) UNITAS 2005

HSL-43 Battlecats

HSL-43 Battle Cats

HSL-43 USS *Lake Champlain* (CG-57) Surge 2007

HSL-43 Battlecats (Subdued)

HSL-43 H-60 Plane Captain

HSL-43 Established 1984, 25th Anniversary

HSL-43 Battlecats (*Gray*)

HSL-43 USS *Thach* (FFG-43) Joint Inter-Agency Task Force-South 2006

HSL-43 Det. 1 USS *Bunker Hill* (CG-52) Lazing around the Gulf

HSL-43 Det. 1 Guardians of Paradise

HSL-43 Det. 2 Sierra USS *Curts* (FFG-38), USS *Bunker Hill* (CG-52), USS *Kansas City* (AOR-3), USS *Leftwich* (DD-984), USS *Missouri* (BB-63), USS *Long Beach* (CGN-9), USS *Hoel* (DDG-13)

HSL-43 Det. 1 USS *Ford* (FFG-54) *Ford Rangers*

HSL-43 Det. 2B USS *Curts* (FFG-38)

HSL-43 Det. 1 USS *Mobile Bay* (CG-53) Mobsters

HM-15 Blackhawks, Sikorsky MH-53 Buno. 162532, Tail code TB, Modex 12, with sled, conducting mine sweeping exercise.

HSL-43 Det. 2 Orient Express

HSL-43 Det. 2 USS *Rodney M. Davis* (FFG-60) Loungers

HSL-43 Det. 2 USS *Kinkaid* (DD-965) Duck Dodgers

HSL-43 Det. 2 USS *Rodney M. Davis* (FFG-60) Loungers (Subdued)

HSL-43 Det. 2 The Nihilists

HSL-43 Det. 2 Strike Force 1987-1988

HSL-43 Det. 2 USS *Jarrett* (FFG-33) Predators

HSL-43 Det. 3 USS *Gary* (FFG-51) Middle East Force 1995

HSL-43 Det. 3 USS *Sampson* (DDG-10) Smokin Aces

HSL-43 Det. 3 USS *Howard* (DDG-82), USS *Paul E. Foster* (DD-964), PCU *Shoup* (DDG-86)

HSL-43 Det. 3 USS *Rodney M. Davis* (FFG-60) Southpac 2001

HSL-43 Det. 3 USS *Sampson* (DDG-10) Smokin Aces (Desert)

HSL-43 Det. 3 USS *Antietam* (CG-54) Westpac 1988–1989

HSL-43 Det. 3 USS *Jarrett* (FFG-33) Carat 2007

HSL-43 Det. 3 USS *Preble* (DDG-88) Westpac 2011 (Desert)

SL-43 Det. 4 USS *Stockdale* (DDG-106)
Return with Honor

HSL-43 Det. 4 Luchadores

HSL-43 Det. 5 "A" Team USS *Paul F. Foster* (DD-964)

SL-43 Det. 4 USS *Antietam* (CG-54)
Jokers 2001

HSL-43 Det. 4 Narco Ninjas USS *Rodney M. Davis* (FFG-60) 2010 (Desert)

HSL-43 Det. 5 USS *Jarrett* (FFG-33) Pacific & Middle East Force 1996

SL-43 Det. 4 USS *Thach*
(FFG-43) Juggernauts

HSL-43 Det. 5 Persian Gulf Escort Service 1987

HSL-43 Det. 5 USS *McCampbell* (DDG-85)
Southpac 2003

HSL-43 Det. 5 USS *Valley Forge* (CG-50) Counter Drug Operations 2004

HSL-43 Det. 6 Warbirds USS *Elliot* (DD-967)

HSL-43 Det. 6 USS *Howard* (DDG-83) Westpac 2008 (Desert)

HSL-43 Det. 5 USS *Chancellorsville* (CG-62)

HSL-43 Det. 6 USS *Ingraham* (FFG-61) Middle East Force 1999

HSL-43 Det. 6 Gearheads USS *Howard* (DDG-83)(Subdued)

HSL-43 Det. 6 Trackers USS *Rentz* (FFG-46)

HSL-43 Det. 6 USS *Howard* (DDG-83) Westpac 2008 (Subdued)

HSL-43 Det. 6 Gearheads USS *Howard* (DDG-83) (Desert)

HSL-43 Det. 7 USS *Elliot* (DD-967) Westpac 1990

HSL-43 Det. 7 Warrior Legacy USS *Jason Dunham* (DDG-109)

HSL-43 Det. 9 USS *Jarrett* (FFG-33) Middle East Force 2000

HSL-43 Det. 7 USS *Ingraham* (FFG-61)

HSL-43 Det. 8 Double Trouble USS *Valley Forge* (CG-50)

HSL-43 Det. 9 Stray Cats

HSL-43 Det. 7 USS *Curts* (FFG-38) Millennium Southpac

HSL-43 Det. 8 Fighting Hell Fish USS *Elliot* (DD-967) 2000

HSL-43 Det. 10 USS *Simpson* (FFG-56) Persian Gulf April 89, Middle East Force January 1990

HSL-43 Det. 10 USS *Cushing* (DD-905) 1990

HSL-43 Det. 10 Wildcats USS *Gary* (FFG-51)

HSL-44 Swamp Fox

HSL-43 Det. 10 USS *Princeton* (CG-59) Lords of the Jungle

HSL-43 Det. 10 Kingpins USS *De Wert* (FFG-45)

HSL-44 Swamp Fox (Subdued)

HSL-43 Det. 10 Westpac 1993

HSL-43 Det. 11 Operation Desert Storm

HSL-44 Swamp Fox (Desert)

HSL-44 Swamp Fox

HSL-44 Battle Readiness E Award 1989

HSL-44 Back to Back Battle Readiness "E" Awards 1995 & 1996

HSL-44 Swamp Fox

HSL-44 Desert Fox, Saddam Don't Surf

HSL-44 Battle Readiness "E" Award 1998

HSL-44 SH-60B Plane Captain

HSL-44 Battle Readiness "E" Award 1995

HSL-44 Golden Wrench Award for Maintenance Excellence 2000 & 2001

HSL-44 Golden Wrench Award for Maintenance Excellence 2000 & 2001

HSL-44 Mission Commander No *Mercy* Lamp's MK III (Silver)

HSL-44 Mission Commander No *Mercy* Lamp's MK III (Gold)

HSL-44 Det. 1 Lumber Jacks

HSL-44 Det. 1 USS *Stump* (DD-978) Middle East Force January 1993

HSL-44 Det. 1 USS *Stump* (DD-978) UNITAS 1994

HSL-44 Det. 1 USS *Conolly* (DD-979)

HSL-44 Det. 1 USS *Briscoe* (DD-977) All Stars 2002–2003

HSL-44 Det. 1 USS *Underwood* (FFG-36) 2004

HSL-44 Det. 1 USS *Vella Gulf* (CG-72) Baltic Operations 2006

HSL-44 Det. 1 USS *Hawes* (FFG-53) A Hawesome Pair, Lamps

HSL-44 Det. 2 USS *Deyo* (DD-989) Med. March 1994

HSL-44 Det. 1 USS *Vella Gulf* (CG-72) 2007

HSL-44 Det. 1 Gladiators

HSL-44 Det. 2 USS *Simpson* (FFG-56)

HSL-44 Det. 1

HSL-44 Det. 1 One Man Wolfpack

HSL-44 Det. 2 USS *De Wert* (FFG-45) Swamp Spuds

HSL-44 Det. 2 USS *Monterey* (CG-61) Hooligans

HSL-44 Det. 2 USS *Deyo* (DD-989)

HSL-44 Det. 2 Trojans USS *Taylor* (FFG-50) 2006–2007

HSL-44 Det. 3 USS *Elrod* (FFG-55) Middle East Force April 1987

HSL-44 Det. 3 USS *Elrod* (FFG-55) Middle East Tour 1991–1992

HSL-44 Det. 3 USS *Elrod* (FFG-55) Standard Naval Force Atlantic February 1993

HSL-44 Det. 3 USS *Hue City* (CG-66) Med. February 1997

HSL-44 Det. 3 USS *Nicholson* (DD-958) Med. 1999

HSL-44 Det. 3 Stingers

HSL-44 Det. 3 USS *Carr* (FFG-52) 2006

HSL-44 Det. 4 USS *Halyburton* (FFG-40) January 2003

HSL-44 Det. 4 USS *Carr* (FFG-52) Red Sea-Med. January 1994

HSL-44 Det. 3 USS *Mason* (DDG-87) Saviors

HSL-44 Det. 5 USS *Carr* (FFG-52) Persian Gulf 1990

HSL-44 Det. 4 Med. 2001

HSL-44 Det. 4 USS *Peterson* (DD-969) Patriots

HSL-44 Det. 4 USS *Carr* (FFG-52) Middle East Force-Red Sea-Maritime Interdiction Force January 1992

HSL-44 Det. 4 USS *Monterey* (CG-61) Med. & Middle East Force 2004-2005

HSL-44 Det. 4 USS *Stephen W. Groves* (FFG-29) Middle East Force March 1996

HSL-44 Det. 5 Alert 7-Auto Dog

HSL-44 Det. 5 USS *Briscoe* (DD-977) Middle East Force March 1992 (Subdued)

HSL-44 Det. 4 USS *Elrod* (FFG-55)

HSL-44 Det. 5 USS *Samuel B. Roberts* (FFG-58) Red Sea-Operation Desert Shield 1990

HSL-44 Det. 5 USS *De Wert* (FFG-45) Counter Drug Operations 2004

HSL-44 Det. 4 USS *Boone* (FFG-28) UNITAS 2011

HSL-44 Det. 5 USS *Briscoe* (DD-977) Middle East Force March 1992

HSL-44 Det. 5 USS *De Wert* (FFG-45) Heart Breakers

HSL-44 Det. 5 USS *Halyburton* (FFG-40) UNITAS 2006

HSL-44 Det. 5 USS *Stephen W. Groves* (FFG-29) Counter Narcotics Operations 2007–2008

HSL-44 Det. 5 USS *Stephen W. Groves* (FFG-29) Counter Narcotics Operations 2007–2008 (Subdued)

HSL-44 Det. 5 Bullet Air

HM-14 H-53's on USS *New Orleans* (LPH-1).

HSL-44 Det. 5 Dangerous Men

HSL-44 Det. 6 USS *Doyle* (FFG-39) UNITAS 39

HSL-44 Det. 5

HSL-44 Det. 6 USS *Elrod* (FFG-55) Standard Naval Force Med.

HSL-44 Det. 5 Playboys

HSL-44 Det. 6 USS *Halyburton* (FFG-40) Med. January 1993

HSL-44 Det. 6 USS *Vella Gulf* (CG-72) Middle East Force March 1995

HSL-44 Det. 6 Fine Malt Lamps

HSL-44 Det. 7 USS *Stephen W. Groves* (FFG-29

HSL-44 Det. 6 USS *John Hancock* (DD-981) Med. & Middle East Force 1999

HSL-44 Det. 6 USS *Bataan* (LHD-5) Strike Group 2007

HSL-44 Det. 7 Hellions USS *Elrod* (FFG-55)

HSL-44 Det. 6 USS *Winston S. Churchill* (DDG-81) 2009–2010

HSL-44 Det. 6 USS *Anzio* (CG-68) Operation Joint Warrior 2008

HSL-44 Det. 7 USS *Normandy* (CG-60) Operation Desert Shield February 1991

HSL-44 Det. 7 USS *Normandy* (CG-60) Operation Desert Storm 1991

HSL-44 Det. 7 USS *Stephen W. Groves* (FFG-29) Med. February 1992

HSL-44 Det. 7 USS *Stephen W. Groves* (FFG-29) Middle East Force February 1994

HSL-44 Det. 7 USS *Nicholson* (DD-982) Med. & Middle East Force 2001-2002

HSL-44 Det. 7 USS *Samuel B. Roberts* (FFG-58) Counter Drug Operations 2003–2004

HSL-44 Det. 8 USS *Nicholas* (FFG-47) Operation Desert Storm & Middle East Force April 1990

HSL-44 Det. 8 USS *Nicholas* (FFG-47) Red Sea & Med. February 1993

HSL-44 Det. 8 USS *Halyburton* (FFG-40) Middle East Force January 1995

HSL-44 Det. 8 USS *Cape St. George* (CG-71) Med. & Middle East Force 2000-2001

HSL-44 Det. 8 USS *John F. Kennedy* (CV-76) Battle Group 2002

HSL-42 Det. 8 USS *Winston S. Churchill* (DDG-81) One Last Ride

HSL-44 Det. 9 Hellcats

HSL-44 Det. 8 USS *George Washington* (CVN-73) Surge Group 2004

HSL-42 Det. 8 USS *Winston S. Churchill* (DDG-81) One Last Ride (Subdued)

HSL-44 Det. 9 Nasty Nine

HSL-44 Det. 8 USS *San Jacinto* (CG-56) Screamin' Boheekas

HSL-42 Det. 8 USS *Winston S. Churchill* (DDG-81) One Last Ride (Desert)

HSL-44 Det. 9 USS *Thomas S. Gates* (CG-51) Med. February 1992

HSL-44 Det. 9 USS *Thomas S. Gates* (CG-51) Med. February 1992

HSL-44 Det. 9 USS *Enterprise* (CVN-65) Carrier Surge Group 2007

HSL-44 Det. 10 USS *Vicksburg* (CG-69) Med. January 1994

HSL-44 Det. 9 USS *Hayler* (DD-997) Middle East Force February 1998

HSL-44 Det. 10 Renegades

HSL-44 Det. 10 USS *Vicksburg* (CG-69) Med. & Middle East Force 1997

HSL-44 Det. 9 Shakas

HSL-44 Det. 10 USS *Samuel B. Roberts* (FFG-58) Persian Gulf November 1992

HSL-44 Det. 10 USS *Philippine Sea* (CG-58) Med. & Middle East Force 1999

HSL-44 Det. 10 USS *Caron* (DD-970) Standard Naval Force Med. 2000–2002

HSL-44 Det. 10 Los Diablos

HSL-45 Wolfpack

HSL-44 Det. 10 USS *Simpson* (FFG-56) Standard Naval Force Atlantic 2004

HSL-44 Det. 10 USS *Forrest Sherman* (DDG-98) Partnership of the Americas 2008

HSL-45 Wolfpack (Subdued)

HSL-44 Det. 10 USS *Halyburton* (FFG-40) Counter Narcotics Operations 2007

HSL-44 Det. 10 USS *Simpson* (FFG-56) Counter Narcotics Terror Operations 2009

HSL-45 Wolfpack (Desert)

HSL-45 Westpac 1990

HSL-45 Det. 1 USS *Jarrett* (FFG-33) 3 November 1992–03 May 1993

HSL-45 Det. 1 USS *Howard* (DDG-83) Westpac 2004

HSL-45 Wolfpack Twenty-Year Anniversary October 03 1986–October 03 2006

HSL-45 Det. 1 USS *Ford* (FFG-54) Westpac 1996

HSL-45 Det. 1 USS *Jarrett* (FFG-33) Carat 2008

HSL-45 Det. 1 USS *Antietam* (CG-54)

HSL-45 Det. 1 USS *Antietam* (CG-54)

HSL-45 Det. 1 USS *Rodney M. Davis* (FFG-60)

HSL-45 Det. 2 USS *Jarrett* (FFG-33)
Iraq-No-Phobia

HSL-45 Det. 2 USS *Oldendorf* (DD-972) & USS *Fletcher* (DD-992) Seaswap 2003

HSL-45 Det. 2 USS *Vandegrift* (FFG-48)
Det Dos Equis 2008

HSL-45 Det. 2 USS *Antietam* (CG-54)
February -August 1994

HSL-45 Det. 2 USS *Halsey* (DDG-97)

HSL-45 Det. 2 USS *Gary* (FFG-51)
Southpac 2009

HSL-45 Det. 2 USS *Ford* (FFG-54)

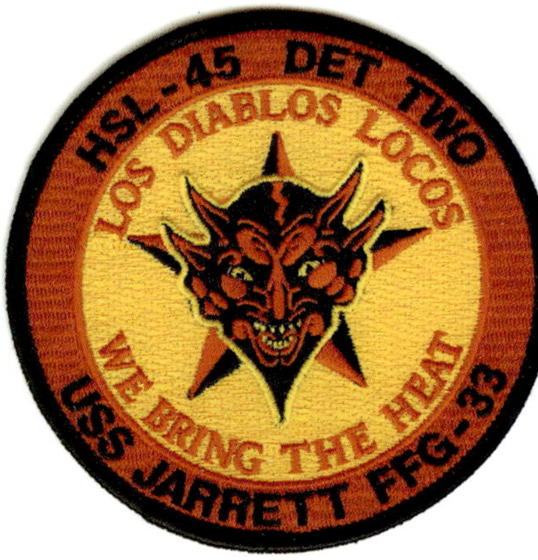

HSL-45 Det. 2 USS *Jarrett* (FFG-33)
Los Diablos Locos

HSL-45 Det. 2 USS *Princeton* (CG-50)
Westpac 2010

HSL-45 Det. 2 USS *Princeton* (CG-59) Westpac 2010 (Subdued)

HSL-45 Det. 3 USS *Oldendorf* (DD-972) Stopping the 100% Pure Colombian

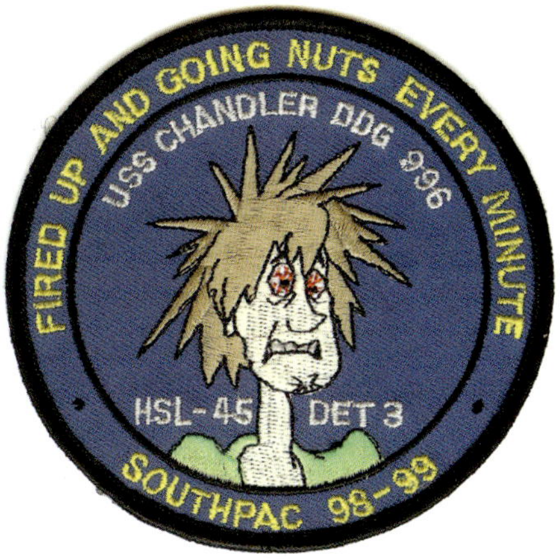

HSL-45 Det. 3 USS *Chandler* (DD-996) Southpac 1998–1999

HSL-45 Det. 2 USS *Princeton* (CG-59) Westpac 2010 (Desert)

HSL-45 Det. 3 USS *De Wert* (FFG-45) Drug War

HSL-45 Det. 3 USS *Rodney M. Davis* (FFG-60) Ninja Monkeys

HSL-45 Det. 3 Spuds Escort Service 1987–1988

HSL-45 Det. 3 Imported San Diego

HSL-45 Det. 3 USS *McClusky* (FFG-41) Rum Runners

HSL-45 Det. 3 USS *McClusky* (FFG-41) 2009–2010

HSL-45 Det. 4 USS *McCampbell* (DDG-85) Commissioning

HSL-45 Det. 4 USS *Cowpens* (CG-63) Westpac 1992–1993

HSL-45 Det. 3 USS *McClusky* (FFG-41) 2009–2010 (Subdued)

HSL-45 Det. 4 Middle East Force 1987–1988

HSL-45 Det. 4 *John Young* (DD-973)

HSL-45 Det. 3 USS *McClusky* (FFG-41) 2009–2010 (Desert)

HSL-45 Det. 4 USS *Cowpens* (CG-63) March 1992

HSL-45 Det. 4 USS *Lake Champlain* (CG-57)

HSL-45 Det. 4 USS *Lake Champlain* (CG-57) (Subdued)

HSL-45 Det. 5 USS *Mustin* (DDG-89)

HSL-45 Det. 5 USS *Bunker Hill* (CG-52) Twenty Year Anniversary

HSL-45 Det. 5 USS *Vincennes* (CG-49) Westpac August–December 1991

HSL-45 Det. 5 USS *Mustin* (DDG-89) (Desert)

HSL-45 Det. 5 USS *Cape St. George* (CG-71) Shooters

HSL-45 Det. 5 Westpack 2000

HSL-45 Det. 5 Beanskinners from Hell (Subdued)

HSL-45 Det. 5 USS *Curts* (FFG-38) 2010

HSL-45 Det. 6 Killer Ducks

HSL-45 Det. 6 USS *Fife* (DD-991) Mission Accomplished

HSL-45 Det. 6 USS *David R. Ray* (DD-971) Flying Wired

HSL-45 Det. 7 USS *Fife* (DD-991) Westpac 1992

HSL-45 Det. 7 USS *Bunker Hill* (CG-52)

HSL-45 Det. 7 USS *Oldendorf* (DD-972)

HSL-45 Det. 7 USS *Ingraham* (FFG-61) January 2003–?

HSL-45 Det. 7 USS *Preble* (DDG-88) It's Not the Size of the Can

HSL-45 Det. 7 USS *Howard* (DDG-83) Westpac 2004

HSL-46 Battle Ready "E" Award 2003

HSL-46 Det. 1 USS *Winston S. Churchill* (DDG-81) Junk Yard Dawgs

HSL-46 Det. 1 Ponies USS *Klakring* (FFG-42) Med. March 1989

HSL-46 Det. 1 USS *Thorn* (DD-988) Med. 2001–2002

Two HSC-28 Dragon Whales, Sikorsky SH-60 Seahawks carry cargo to USS *Dwight D. Eisenhower* (CVN-69) from USNS *Arctic* (T-AOE-8), January 25, 2007.

HSL-46 Det. 1 USS *O'Bannon* (DD-987) Middle East Force March 1997

HSL-46 Det. 1 USS *Kauffman* (FFG-59)

HSL-46 Det. 1 USS *De Wert* (FFG-45) The Dukes of Mayport 2008 (Desert)

HSL-46 Det. 2 USS *Underwood* (FFG-36) Middle East Force February 1989

HSL-46 Det. 2 USS *Oscar Austin* (DDG-79) Tail Chasers

HSL-46 Det. 2 USS *Spruance* (DD-963) Morgan's Last Call

HSL-46 Det. 2 USS *Spruance* (DD-963) Morgan's Last Call (Desert)

HSL-46 Det. 2 USS *Peterson* (DD-969) Middle East Force February 1993

HSL-46 Det. 2 USS *Roosevelt* (DDG-80) Money Makers 2006

HSL-46 Det. 2 You Have Been Punished

HSL-46 Det. 2 (Subdued)

HSL-46 Det. 2 (Desert)

HSL-46 Det. 2 Chupacabras USS *Halyburton* (FFG-40) USS *Eisenhower* (CGN-69) Carrier Strike Group 2009

HSL-46 Det. 2 Chupacabras USS *Halyburton* (FFG-40) USS *Eisenhower* (CGN-69) Carrier Strike Group 2009 (Desert)

HSL-46 Det. 3 USS *Boone* (FFG-28) Middle East Force March 1989

HSL-46 Det. 3 USS *Underwood* (FFG-36)

HSL-46 Det. 3 USS *Comte De Grasse* (DD-974) Med. February 1992

HSL-46 Det. 3 USS *Robert G. Bradley* (FFG-49) Med. February 1996

HSL-46 Det. 3 USS *Hayler* (DD-997) Med. & Middle East Force 2001–2002

HSL-46 Det. 3 USS *Doyle* (FFG-39) Taking Care of Business

HSL-46 Det. 3 USS *Doyle* (FFG-39) Taking Care of Business

HSL-46 Det. 3 USS *McInerney* (FFG-8) Counter Narcotics Operations

HSL-46 Det. 3 USS *Carr* (FFG-52) Cutthroats

HSL-46 Det. 3 USS *Normandy* (CG-60) Screamin' Seagulls

HSL-46 Det. 4 USS *Spruance* (DD-963) Med. 2002

HSL-46 Det. 4 USS *Briscoe* (DD-977) Middle East Force 1998

HSL-46 Det. 4 USS *John L. Hall* (FFG-32) Med. 1990

HSL-46 Det. 4 USS *Spruance* (DD-963) Med. 2002

HSL-46 Det. 4 USS *Vicksburg* (CG-69)

HSL-46 Det. 4 USS *Anzio* (CG-68) Players

HSL-46 Det. 4 USS *Nicholas* Mystery Cruise 2003

HSL-46 Det. 4 USS *Vicksburg* (CG-69) (Desert)

HSL-46 Det. 4 USS *Anzio* (CG-68) Players (Subdued)

HSL-46 Det. 4 USS *Mason* (DDG-87) *Eisenhower* (CVN-69) Strike Group 2006–2007

HSL-46 Det. 4-249th Cutlass Strike Force

HSL-46 Det. 4 USS *Anzio* (CG-68) Players (Desert)

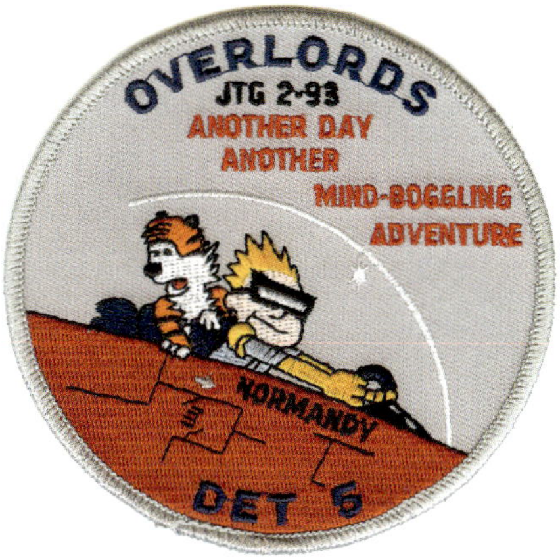

HSL-46 Det. 5 Caribbean Studs

HSL-46 Det. 5 USS *Normandy* (CG-60) Joint Task Group 2 1993

HSL-46 Det. 5 USS *Taylor* (FFG-50) Med. 1997

HSL-46 Det. 5 USS *Gettysburg* (CG-54) Gladiators

HSL-46 Det. 5 USS *Anzio* (CG-68) Med. & Middle East Force 1994-1995

HSL-46 Det. 5 USS *John L. Hall* (FFG-32) Counter Drugs Operations 2004

HSL-46 Det. 5 USS *Vella Gulf* CG-72) Seabats

HSL-46 Det. 5 USS *Anzio* (CG-68) Med. & Gulf Cruise 1994–1995

HSL-46 Det. 5 USS *Gettysburg* (CG-64) Counter Narcotics Operations 2005–2006

HSL-46 Det. 5 USS *Philippine Sea* (CG-58) & USS *Normandy* (CG-60) Spiderpigs

HSL-46 Det. 5 USS *Enterprise* (CVN-65) Strike Group 2008

HSL-46 Det. 5 USS *Carr* (FFG-52) Hell Hounds

HSL-46 Det. 5 USS *Carr* (FFG-52) Hell Hounds (Subdued)

HSL-46 Det. 5 USS *Carr* (FFG-52) Hell Hounds (Desert)

HSL-46 Det. 6 USS *O'Bannon* (DD-987) UNITAS 40-1999

HSL-46 Det. 6 USS *O'Bannon* (DD-987)

HSL-46 Det. 6 Arthur W. Radford (DD-968) Six Shooters

HSL-46 Det. 6 USS *Robert G. Bradley* (FFG-49) Counter Drug Operations & UNITAS 1983

HSL-46 Det. 6 Agents of the Free 1991–1992

HSL-46 Det. 6 USS *Vella Gulf* (CG-72) Machines

HSL-46 Det. 7 USS *Philippine Sea* (CG-58) Med. January 1994

HSL-46 Det. 6 USS *Kauffman* (FFG-59) Wildcards

HSL-46 Det. 6 USS *Vicksburg* (CG-69) Big Irons

HSL-46 Det. 7 USS *Normandy* (CG-60) What Are You Going to Do Tonight Brain?

HSL-46 Det. 6 USS *Robert G. Bradley* (FFG-49) UNITAS 2003

HSL-46 Det. 6 USS *Vicksburg* (CG-69) Big Irons (Desert)

HSL-46 Det. 7 USS *Normandy* (CG-60) Nomads (Desert)

HSL-46 Det. 7 Warhawks USS *Philippine Sea* (CG-58) Med. March 1990

HSL-46 Det. 7 Super Troopers (Desert)

HSL-46 Det. 7 USS *Klakring* (FFG-42) Standard Naval Force Atlantic 2005

HSL-46 Det. 7 Archangels

HSL-46 Det. 7 USS *Simpson* (FFG-56) Counter Narcotics Operations

HSL-46 Det. 8 USS *Winston S. Churchill* (DDG-81) Bulldogs

HSL-46 Det. 7 Super Troopers

HSL-46 Det. 7 USS *Klakring* (FFG-42) Standard Naval Force Atlantic 2004

HSL-46 Det. 8 USS *Doyle* (FFG-39) Rude Dogs

HSL-46 Det. 8 USS *Nicholson* (DD-982) Middle East Force 1997

HSL-46 Det. 8 USS *Stephen W. Groves* (FFG-29) Counter Narcotics Operations 2006

HSL-46 Det. 9 USS *Monterey* (CG-61) Med. January 1992

HSL-46 Det. 8 USS *De Wert* (FFG-45) Standard Naval Force Atlantic & Standard Naval Force Med. February 2002

HSL-46 Det. 8 USS *Bulkeley* (DDG-84) Hitmen

HSL-46 Det. 9 USS *Monterey* (CG-61) Med. January 1992

HSL-46 Det. 8 Wigsplitters

HSL-46 Det. 9 USS *Monterey* (CG-61) 1991–1992

HSL-46 Super Det. 9 USS *Ticonderoga* (CG-47) Counter Drug Operations 1998

HSL-46 Det. 9 USS *Moosbrugger* (DD-980) Standard Naval Force Atlantic 2000

HSL-46 Det. 9 USS *Anzio* (CG-68) USS *Eisenhower* (CVN-69) Strike Group 2006–2007

HSL-46 Det. 9 The Wreckin Crew (Desert)

HSL-46 Det. 9 USS *Taylor* (FFG-50) Standard Naval Force Med. Summer 2002

HSL-46 Det. 9 USS *Anzio* (CG-68) USS *Eisenhower* (CVN-69) Strike Group 2006–2007

HSL-46 Det. 10 USS *Robert G. Bradley* (FFG-49) Icemen

HSL-46 Det. 9 USS *Anzio* (CG-68) USS *Eisenhower* (CVN-69) Strike Group 2006–2007

HSL-46 Det. 9 The Wreckin Crew

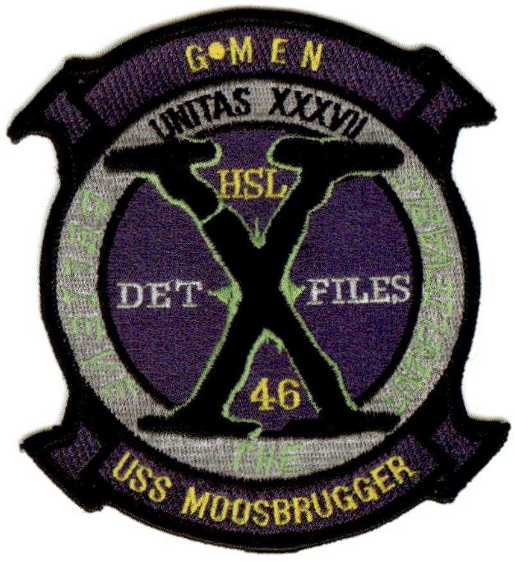

HSL-46 Det. 10 USS *Moosbrugger* (DD-980) UNITAS 37

HSL-46 Det. 10 USS *Doyle* (FFG-39) Weedwackers

HSL-47 Saberhawks

HSL-47 Saberhawks SH-60B (Desert)

HSL-47 Saberhawks

HSL-47 Saberhawks (Desert)

HSL-47 Average J.O.S.

HSL-47 Saberhawks

HSL-47 Sabrehawks SH-60B Plane Captain

HSL-47 Night Check Zombie Squad

HSL-47 Westpac 2008 Last SH-60B Ride

HSL-47 Det. 1 USS *McClusky* (FFG-31) Carat 1997

HSL-47 Det. 1 Bravo Bonzai Bandits

HSL-47 Det. 1C 1989-1990

HSL-47 Det. 1D Pacific Exercise 1989- Persian Excursion 1990

HSL-47 Det. 2 USS *Antietam* (CG-54)

HSL-47 Mini Det. 2 USS *John Young* (DD-973)

HSL-47 Det. 2

X-21 Blackjack Sikorski SH-60B Seahawk anding on the USS *Sea Fighter* (FSF-1, an luminum catamaran) during deck ertifications off San Diego.

HSL-47 Det. 2 USS *Valley Forge* (CG-50) Operation Desert Shield 1991

HSL-47 Det. 2 USS *Kinkaid* (DD-965) Westpac 1997–1998

HSL-47 Det. 2 All Stars USS *Shoup* (DDG-86)
Westpac 2008

HSL-47 Det. 3 USS *Ford* (FFG-54)

HSL-47 Det. 3 USS *Momsen* (DDG-92)
Shockers (Subdued)

HSL-47 Det. B2C USS *Abraham Lincoln*
(CGN-72) Strike Group

HSL-47 Det. 3 USS *Rentz* (FFG-46)
Westpac 1995

HSL-47 Det. 3 USS *Momsen* (DDG-92)
Shockers (Desert)

HSL-47 Det. 3 USS *David R. Ray* (DD-971)
Maritime Interdiction Force
Summer 1992

HSL-47 Det. 3 USS *Shiloh* (CG-67) Persian Gulf
& Pacific 1998

HSL-47 Det. 4 USS *Chancellorsville* (CG-62)
Hotel California

SL-47 Det. 4 USS *Curts* (FFG-38)

HSL-47 Det. 5 USS *Oldendorf* (DD-972) Rimpac & Westpac 1998–1999

HSL-47 Det. 6 Delivers

SL-47 Det. 4 USS *Constellation* (CV-64) Westpac 2001

HSL-47 Det. 5 Terminate with Extreme Prejudice

HSL-47 Det. 6 Keep Pushin'

SL-47 Det. 5

HSL-47 Det. 5 USS *Princeton* (CG-59) Westpac 2003

HSL-47 Det. 6 USS *Ford* (FFG-54)-USS *John Young* (DD-973)-USS *Russell* (DDG-59) Middle East Force Exercise

HSL-47 Det. 6 Westpac-Persian Gulf 1991

HSL-47 Det. 6 USS *Doyle* (FFG-39) Caribbean 2000

HSL-47 Det. 8 USS *Princeton* (LPH-5) Operation Desert Storm 1990–1991

HSL-47 Det. 9 Westpac August–December 1990

HSL-47 Det. 9 Middle East Force 1997

HSL-47 Det. 18 Swinging Ducks

HSL-48 Vipers

HSL-48 Vipers (Subdued)

HSL-48 Vipers (Desert)

SL-48 SH-60B Plane Captain

SL-48 Twenty Year Anniversary 1989–2009

SL-48 Det. 1 USS *Thomas S. Gates* (CG-51) Renegades

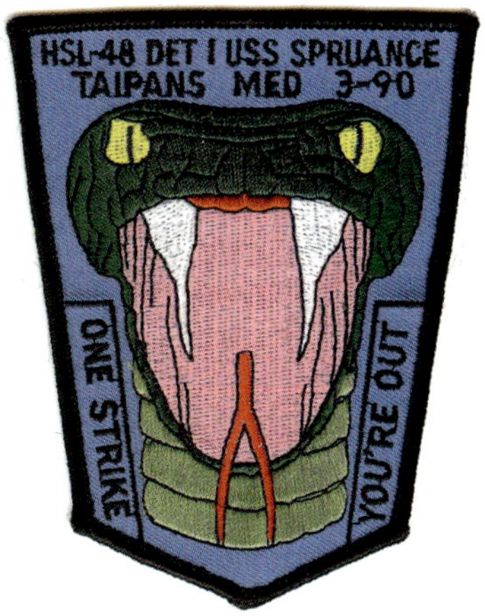

HSL-48 Det. 1 USS *Spruance* (DD-963) Taipans & Med. March 1990

HSL-48 Det. 1 USS *Spruance* (DD-963) Middle East Force March 1993

HSL-48 Det. 1 USS *Doyle* (FFG-39) Standard Naval Force Atlantic- World Tour 2003

HSL-48 Det. 1 USS *Forrest Sherman* (DDG-98) Original Gangsters

HSL-48 Det. 1 USS *Forrest Sherman* (DDG-98) Original Gangsters (Subdued)

HSL-48 Det. 2 USS *John L. Hall* (FFG-32) Counter Drug Operations

HSL-48 Det. 2 USS *Hawes* (FFG-53) Operation Desert Shield

HSL-48 Det. 2 USS *McInerney* (FFG-8) UNITAS 35 July-December 1994

HSL-48 Det. 2 USS *John L. Hall* (FFG-32) NAF Mayport Vice- CIT Operations 2011

HSL-48 Det. 2 RA Skulls USS *Hawes* (FFG-53) Med. 1993

HSL-48 Det. 2 UNITAS 35 Los Vaqueros 1994

HSL-48 Det. 2 USS *John C. Stennis* (CVN-74) Surge Cruise 2012–2013

HSL-48 Det. 2 USS *Hawes* (FFG-53) Med. February 1993 Wild Ones

HSL-48 Det. 2 USS *McInerney* (FFG-8) Standard Naval Force Atlantic 1996

HSL-48 Det. 2 North Africa MQ-8B Firescout-Eye in the Sky-Special Operating Forces 2013

HSL-48 Det. 2 USS *Vella Gulf* (CG-72) 2014

HSL-48 Det. 3 USS *Roosevelt* (DDG-80) Big Fluffy Bunnies

HSL-48 Det. 3 Official Det. of the Millennium

HSL-48 Det. 2 USS *Vella Gulf* (CG-72) 2014 (Subdued)

HSL-48 Det. 3 USS *Simpson* (FFG-56) Med. 1998

HSL-48 Det. 3 USS *Saipan* (LHA-2) & USS *Hue City* (CG-66) 2004-2005

HSL-48 Det. 2 USS *Vella Gulf* (CG-72) 2014 (Desert)

HSL-48 Det. 3 USS *Caron* (DD-970) Viper Vengeance

HSL-48 Det. 3 USS *Saipan* (LHA-2) & *Hue City* (CG-66) 2004–2005 (Desert)

HSL-48 Det. 3 USS *Elrod* (FFG-55) Commence Project Mayhem

HSL-48 Det. 3 USS *Elrod* (FFG-55) Commence Project Mayhem (Desert)

HSL-48 Det. 3 USS *Underwood* (FFG-36) Conquistadors SH-60B

HSL-48 Det. 4 USS *Bulkeley* (DDG-84) USS *Enterprise* (CVN-65) Strike Group

HSL-48 Det. 4 USS *Nicholas* (FFG-47) Hustlers

HSL-48 Det. 4 USS *Nicholas* (FFG-47) Hustlers (Desert)

HSL-48 Det. 4 USS *McInerney* (FFG-8) Counte Drug Operations & UNITAS 2001

HSL-48 Det. 4 USS *McInerney* (FFG-8) Counte Drug Operations & UNITAS 2003

HSL-48 Det. 4 USS *McInerney* (FFG-8) Standard Naval Force Atlantic 2005

SL-48 Det. 4 USS *San Jacinto* (CG-56) Flying Tanukis 2007–2008

HSL-48 Det. 5 USS *Boone* (FFG-28) Med. & Red Sea 1993

HSL-48 Det. 5B USS *Leyte Gulf* (CG-55) Misfits

SL-48 Det. 4 USS *San Jacinto* (CG-56) Flying Tanukis 2007–2008 (Desert)

HSL-48 Det. 5 USS *Taylor* (FFG-50) Millennium Cruise

HSL-48 Det. 5 USS *Hawes* (FFG-53) Ever Ready Ever Fearless

SL-48 Det. 5 USS *Klakring* (FFG-42) Middle East Force 1992

HSL-48 Det. 5 USS *Leyte Gulf* (CG-55) Bohica Air

HSL-48 Det. 5 USS *Wasp* (LHD-1) Middle East Force 2004

HSL-48 Det. 5 USS *Boone* (FFG-28) SNMG 2006

HSL-48 Det. 5 USS *Leyte Gulf* (CG-55) Bohica Air (Desert)

HSL-48 Det. 5 USS *Monterey* (CG-61) Murican 2013 (Desert)

HSL-48 Det. 5 USS *Leyte Gulf* (CG-55) Bohica Air

HSL-48 Det. 5 USS *Monterey* (CG-61) 'Muricans 2013

HSL-48 Det. 6 Hooligans

HSL-48 Det. 5 USS *Leyte Gulf* (CG-55) Bohica Air (Subdued)

HSL-48 Det. 5 USS *Monterey* (CG-61) Muricans 2013 (Subdued)

HSL-48 Det. 6 USS *O'Bannon* (DD-987) Forward Deployed

HSL-48 Det. 6 USS *Robert G. Bradley* (FFG-49) Middle East Force February 1992

HSL-48 Det. 6 Dirty Deeds Done Dirt Cheep

HSL-48 Det. 7 Big Guns

HSL-48 Det. 6 USS *Carr* (FFG-52) Middle East Force 1998

HSL-48 Det. 6 USS *Stump* (DD-978) Paleriders

HSL-48 Det. 7 USS *Halyburton* (FFG-40) Counter Drug Operations 2004-2005

HSL-48 Det. 6 USS *Winston S. Churchill* (DDG-81) Dirty Dragons

HSL 48 Det. 7 USS *Comte De Grasse* (DD-974) UNITAS 38 1997

HSL-48 Det. 7 USS *Anzio* (CG-68) Guard Dogs

HSL-48 Det. 7 USS *Anzio* (CG-68)
Guard Dogs (Desert)

HSL-48 Det. 7 USS *Samuel B. Roberts* (FFG-58)
Want a Ride? Summer 2007

HSL-48 Det. 8 USS *San Jacinto* (CG-52) Med.
January 1996

HSL-48 Det. 7 Bone Crushers

HSL-48 Det. 8 Big Nasties

HSL-48 Det. 8 USS *Normandy* (CG-60)
Med. 1998

HSL-48 Det. 7 USS *Farragut* (DDG-99)
Deadly Sinners

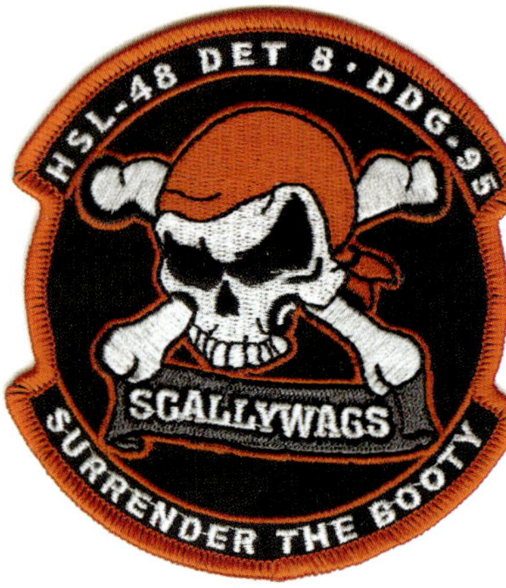

HSL-48 Det. 8 USS *James E Williams*
(DDG 95) Scallywags

HSL-48 Det. 8 USS *Ticonderoga* (CG-47)
Counter Drug Operations & UNITAS 2001

HSL-48 Det. 8 USS *Anzio* (CG-68) Med. & Middle East Force 2003

HSL-48 Det. 8 USS *Nicholas* (FFG-47) USS *Enterprise* (CVN-65) Carrier Strike Group 2006

HSL-48 Det. 9 USS *Philippine Sea* (CG-58) Med. & Middle East Force 2001

HSL-48 Det. 8 USS *Anzio* (CG-68) Med. & Middle East Force 2003 (Subdued)

HSL-48 Det. 9 UNITAS 34 God Bless America

HSL-48 Det. 9 USS *Boone* (FFG-28) Warriors of Oz

HSL-48 Det. 8 USS *Nicholas* (FFG-47) USS *Enterprise* (CVN-65) Carrier Strike Group 2006

HSL-48 Det. 9 USS *Roosevelt* (DDG-80)

HSL-48 Det. 9 USS *Hayler* (DD-997) Counter Narcotics Operations 2003

HSL-48 Det. 9 USS *De Wert* (FFG-45) Counter Narcotics Operations 2005-2006

HSL-48 Det. 9E USS *John L. Hall* (FFG-32) Regulators 2009–2010 (Desert)

HSL-48 Det. 9 USS *Taylor* (FFG-50) Barefoot Bandits (Desert)

HSL-48 Det. 9E Ready-Reserve 2009

HSL-48 Det. 9 USS *Taylor* (FFG-50) Barefoot Bandits

HSL-48 Det. 10 War Eagles

HSL-48 Det. 9E USS *John L. Hall* (FFG-32) Regulators 2009–2010

HSL-48 Det. 9 USS *Taylor* (FFG-50) Barefoot Bandits (Subdued)

HSL-48 Det. 10 USS *Stump* (DD-978) Middle East Force March 1996

HSL-48 Det. 10 USS *Elwood* (FFG-55) Med. 1998

HSL-48 Det. 10 USS *Anzio* (CG-68) Med. 2000

HSL-48 Det. 10 USS *Thorn* (DD-938) Med. 2003

HSL-48 Det. 10 USS *Hawes* (FFG-53) Counter Drug Operations 2008-2009

HSL-48 Det. 10 USS *Hawes* (FFG-53) Counter Drug Operations 2008-2009 (Desert)

HSL-49 Scorpions

HSL-49 Scorpions

HSL-49 Scorpions (Subdued)

HSL-49 Scorpions (Desert)

HSL-49 Plane Captain

HSL-49 A Team USS *Chancellorsville* (CG-62)

HSL-49 Det. 1 USS *Ford* (FFG-64) Untouchables

HSL-49 SH-60 Outlaw Seahawk

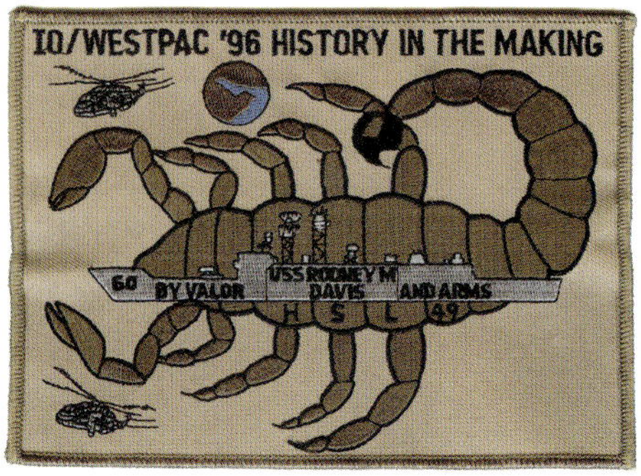

HSL-49 USS *Rodney M Davis* (FFG-60) Indian Ocean, Westpac 1996

HSL-49 Det. 1 USS *Rentz* (FFG-46) Knuckleheads 2009

HSL-49 Last Man Standing

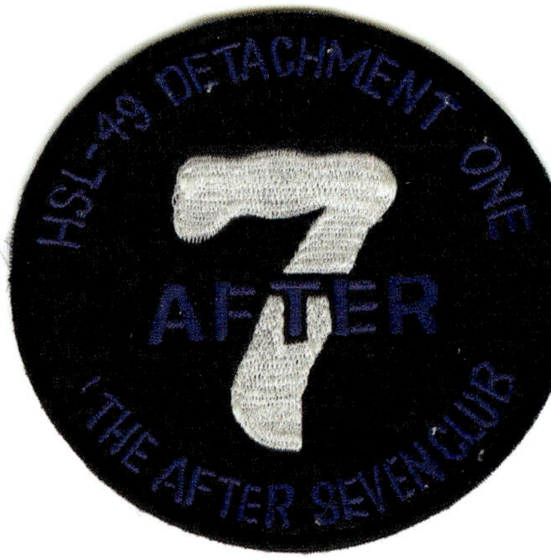

HSL-49 Det. 1 The After Seven Club

HSL-49 Det. 1 USS *Rentz* (FFG-46) Knuckleheads (Subdued) 2009

HSL-49 Det. 1 Dirty Deeds Done Dirt Cheap

HSL-49 Det. 2 USS *Vincennes* (CG-49) Westpac June-December 1994

HSL-49 Det. 2 USS *Curts* (FFG-38) Blunt Force

HSL-49 Det. 2 USS *Curts* (FFG-38) Blunt Force (Desert)

HSL-49 Det. 2 USS *Pinckney* (DDG-91) Outlaw Patch

HSL-49 Det. 2 USS *Pinckney* (DDG-91) The Tenacious Det.

HSL-49 Det. 2 USS *Halsey* (DDG-97) Bull Gators

HSL-49 Det. 2 USS *Halsey* (DDG-97) Bull Gators (Subdued)

HSL-51 Warlords Sikorsky SH-60B Seahawk firing an AGM-119B Penquin missile during aircrew weapon certifications.

HSL-49 Det. 2 USS *Halsey* (DDG-97) Bull Gators (Desert)

HSL-49 Det. 2 USS *Thach* (FFG-43) Hit Squad

HSL-49 Det. 2 USS *Thach* (FFG-43) Hit Squad (Subdued)

HSL-49 Det. 2 USS *Thach* (FFG-43) Hit Squad (Desert)

HSL-49 Det. 2 USS *David R. Ray* (DD-971) Southpac May–November 2001

HSL-49 Det. 2 USS *Rodney M. Davis* (FFG-60) Westpac 2003

HSL-49 Det. 2 USS *Bunker Hill* (CG-52) Band of Brothers 2005

HSL-49 Det. 3 USS *Jarrett* (FFG-33) All Stars

HSL-49 Det. 3 USS *Howard* (DDG-83) Westpac 2006

HSL-49 Det. 2 USS *Princeton* (CG-59) Guardians 2008

HSL-49 Det. 3 USS *Lake Champlain* (CG-57) Westpac Six Months in the Dark 1997–1998

HSL-49 Det. 3 USS *McClusky* (FFG-41) Counte Drug Operations Summer 2008

HSL-49 Det. 2 USS *Ingraham* (FFG-61) CTOC 2014

HSL-49 Det. 3 USS *Antietam* (CG-54) Trans World Rotors 2005

HSL-49 Det. 3 USS *Jarrett* (FFG-33) Southpac 2009

HSL-49 Det. 3 USS *Jarrett* (FFG-33) CIT 2010

HSL-49 Det. 3 USS *Ingraham* (FFG-61) Southpac 2011–2012 (Desert)

HSL-49 Det. 3 USS *Gary* (FFG-51) Counter Narcotics Operations 2012–2013 (Desert)

HSL-49 Det. 3 USS *Ingraham* (FFG-61) Southpac 2011–2012

HSL-49 Det. 3 USS *Gary* (FFG-51) Counter Narcotics Operations 2012–2013

HSL-49 Det. 3 USS *Vandegrift* (FFG-48) Predators 2014

HSL-49 Det. 3 USS *Ingraham* (FFG-61) Southpac 2011–2012 (Subdued)

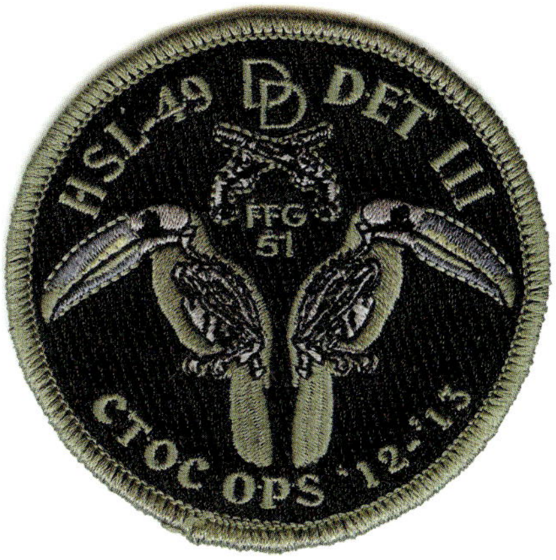

HSL-49 Det. 3 USS *Gary* (FFG-51) Counter Narcotics Operations 2012–2013 (Subdued)

HSL-49 Det. 4 USS *Shiloh* (CG-57) Rimpac 2000-Westpac 2000–2001

HSL-49 Det. 4 USS *Princeton* (LPH-5) Westpac 2007

HSL-49 Det. 4 USS *Chancellorsville* (CG-62) USS *Ronald Reagan* (CVN-76) Surge 2009

HSL-49 Det. 4 USS *Bunker Hill* (CG-52) Dirty Birds

HSL-49 Det. 4 USS *Bunker Hill* (CG-52) Dirty Birds (Subdued)

HSL-49 Det. 4 USS *Bunker Hill* (CG-52) Dirty Birds (Desert)

HSL-49 Det. 4 USS *Bunker Hill* (CG-52) Angry Beavers

HSL-49 Det. 4 USS *Bunker Hill* (CG-52) Angry Beavers (Desert)

HSL-49 Det. 4 USS *Bunker Hill* (CG-52) Angry Beavers

HSL-49 Det. 4 USS *Preble* (DDG-88) Westpac 2013

HSL-49 Det. 4 USS *Preble* (DDG-88) Westpac 2013 (Subdued)

HSL-49 Det. 5 USS *Elliot* (DD-967) Westpac-Persian Gulf July 1993–January 1994

HSL-49 Det. 5 USS *Valley Forge* (CG-50) Westpac 2002–2003

HSL-49 Det. 4 USS *Ingraham* (FFG-61)

HSL-49 Det. 5 USS *Gridley* (DDG-101) Firebirds Never Say Die

HSL-49 Det. 5 USS *McCampbell* (DDG-85) Westpac 2006

HSL-49 Det. 4 USS *Gary* (FFG-51) Rimpac 2014

HSL-49 Det. 5 USS *Gridley* (DDG-101) Firebirds Never Say Die

HSL-49 Det. 5 "Z" Det. It's All About Survival

HSL-49 Det. 5 USS *Pinckney* (DDG-91) Westpac 2007

HSL-49 Det. 5 USS *Thach* (FFG-43) Westpac 2008

HSL-49 Det. 5 USS *Thach* (FFG-43) Scallywags

HSL-49 Det. 5 The Wreckers CTOC 2013

HSL-49 Det. 5 The Wreckers CTOC 2013 (Desert)

HSL 49 Det. 6 USS *Cowpens* (CG-63) Loaded for Bull

HSL-49 Det. 6 Super Heroes (Outlaw Patch)

HSL-49 Det. 6 Super Heroes

HSL-49 Det. 6 Hard Way Café

SL-49 Det. 6 Gunslingers

HSL-49 Det. 6 Immortals USS *Cowpens* (CG-63)

HSL-49 Det. 7 USS *Cowpens* (CG-63) & USS *Mobile Bay* (CG-53) 2000

SL-49 Det. 6 Trojans USS *Gridley* (DDG-101)

HSL-49 Det. 6 USS *Crommelin* (FFG-37) Counter Drug Operations November 2007–May 2008

HSL-49 Det. 7 USS *Cowpens* (CG-63) & USS *Mobile Bay* (CG-53) 2000

SL-49 Det. 6 USS *Shiloh* (CG-67) 2002

HSL-49 Det. 6 USS *Vandegrift* (FFG-48) Carat 2010

HSL-49 Det. 8D USS *Rodney M. Davis* (FFG-60) Pacific February–August 1996

HSL-49 Det. 8 Team 49 Transit August 5, 1997–?

HSL-49 Det. 8L 2000

HSL-51 Warlords

HSL-49 Det. 8I

HSL-49 Det. 9 (GFY) Censored Removed for Content

HSL-51 Warlords (Subdued)

HSL-49 Det. 8J USS *Vandegrift* (FFG-48) Westpac 1999

HSL-49 Det. 10 USS *Chandler* (DDG-996) Det. 10 of the Caribbean

HSL-51 Warlords (Desert)

HSL-51 Warlords (Stealth)

HSL-51 Warlords Cherry Blossoms

HSL-51 Warlords SH-60 Plane Captain

HSL-51 Warlords (Stealth)

HSL-51 Warlords

HSL-51 Safety-Combat Readiness "E" & "A" Awards 1998

HSL-51 Warlords Cherry Blossoms

HSL-51 SH-3B Seahawk

HSL-51 International Champions of Battle "E" & Isbell Awards 2002

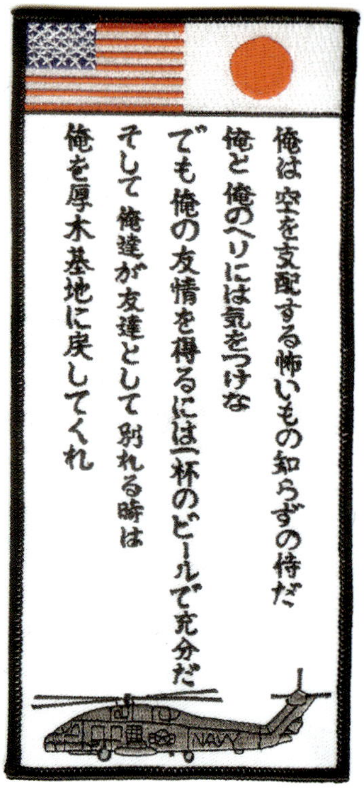

"I am a Warlord fearless master of the sky. No enemy is safe from me and my helicopter. But my friendship is yours for the price of a beer and when we part as friends, please return me to Atsugi Base."

HSL-51 Warlords 20th Anniversary

HSL-51 Cherry Blossom Festival

HSL-51 LAMPS

HSL-51 SH-60 Tail

HSL-51 Blood Patch

HSL-51 Tomodachi

HSL-51 Warlords The Varsity Squad

HSL-51 Det. 1 In the Zone

HSL-51 Det. 1 In the Doghouse Again

HSL-31 Det. 1 Part of the Machine BAMPS

HSL-51 Det. 1 Fight Club

HSL-51 Det. 1 USS *Vandegrift* (FFG-48)
Operation Enduring Freedom & Operation Iraqi Freedom - Soh-Bar Escort 2003

HSL-51 Det. 1 Don't Stop Believin'!

HSL-51 Det. 1 USS *O'Brien* (DD-725)
Det. 1 in Black

HSL-51 Det. 1 One Way

HSL-51 Det. 1 Seadevils

HSL-51 Det. 1 Mobsters

HSL-51 Det. 1 We Got Your Bird

HSL-51 Det. 2 Australia 1999

HSL-51 Det. 1 America's Det.

HSL-51 Det. 1 Bloodhawks

HSL-51 Det. 2 SH-60 Get in the Box

HSL-51 Det. 1 The Short Bus

HSL-51 Det. 2 USS *Cowpens* (CG-63) Just Say When

HSL-51 Det. 2 World Tour 2008

HSL-51 Det. 2 We Got Your Shotgun Right Here!

HSL-51 Det. 2 LAMPS Evolved

HSL-51 Det. 3 You Can Trust Us

HSL-51 Det. 2 Everyone's Fantasy

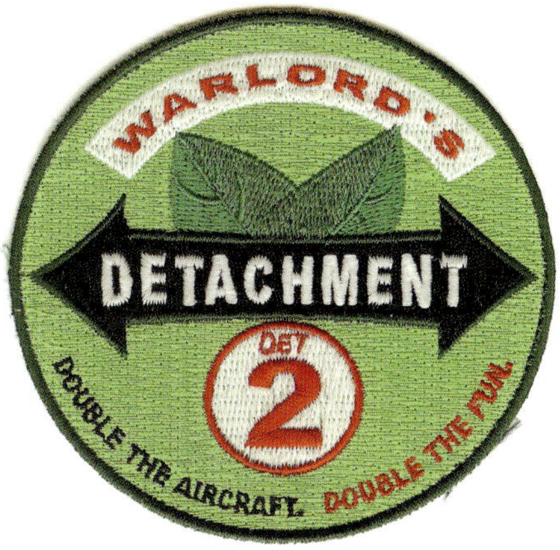

HSL-51 Det. 2 Double the Aircraft, Double the Fun

HSL-51 Det. 3 USS *Kitty Hawk* (CV-63)

HSL-51 Det. 2 USS *Cowpens* (CG-63) Dragon Kings

HSL-51 Det. 3 The Answer is Always 3

HSL-51 Det. 3 Rotor Cowboys

HSL-51 Det. 3 USS *Kitty Hawk* (CV-63) Summer Trip 2006

HSL-51 Det. 3 Far East Choppers

HSL-51 Det. 3

HSL-51 Det. 3 Nothin' But Love

HS-5 Nightdippers Sikorsky SH-60F Seahawk Buno. 164799, Tail code AG, Modex 810 with USS *Dwight D. Eisenhower* (CVN-69), conducting support for Operation Enduring Freedom.

HSL-51 Det. 3 USS *Kitty Hawk* (CV-63) & USS *George Washington* (CVN-73) Black Sheep

HSL-51 Det. 3 USS *Kitty Hawk* (CV-63) & USS *George Washington* (CVN-73) 2004-2011

HSL-51 Det. 4 USS *Chancellorsville* (CG-62) Taking Care of Business

HSL-51 Det. 3 USS *Kitty Hawk* (CV-63) & USS *George Washington* (CVN-73) Black Sheep (Desert)

HSL-51 Det. 4 Shanghi'd

HSL-51 Det. 4

HSL-51 Det. 3 (CVW-5) USS *George Washington* (CVN-73)

HSL-51 Det. 4 Cute and Cuddly Boys

HSL-51 Det. 4 USS *Shiloh* (CG-67) Warm and Fuzzy

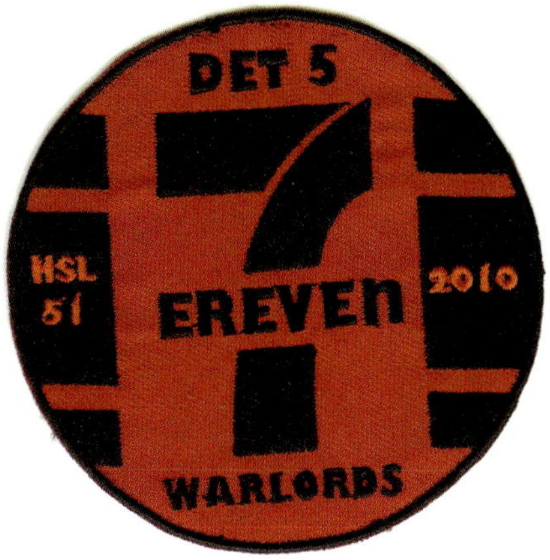

SL-51 Det. 5 Go Ugly Early

HSL-51 Det. 5 Flying Death

HSL-51 Det. 5 2010 Always Up

SL-51 Det. 5 Operation Enduring Freedom & Operation Iraqi Freedom 2003 (Desert)

HSL-51 Det. 5 Rotortrash Cowboys

HSL-51 Det. 5 Warlords 2010

SL-51 LAMPS Det. 5 Okinawa

HSL-51 Det. 5 USS *McCampbell* (DDG-85) Livin it-Lovin it

HSL-51 Det. 5 Busy Being Tactical (Subdued)

HSL-51 Det. 5 Speak Tactical to Me

HSL-51 Det. 6 Give Till It Hurts

HSL-51 Det. 6 We're No Sideshow

HSL-51 Det. 5 Red Hand

HSL-51 Det. 6 It Got Weird Didn't It

HSL-51 Det. 6 Seventh Fleet Sentry

HSL-51 Det. 6 USS *Oldendorf* (DD-972) 1994

HSL-51 Det. 6 Gunslingers

HSL-51 Det. 6 Squeezing the Juice Out of Whatever Homeguard Gives Us

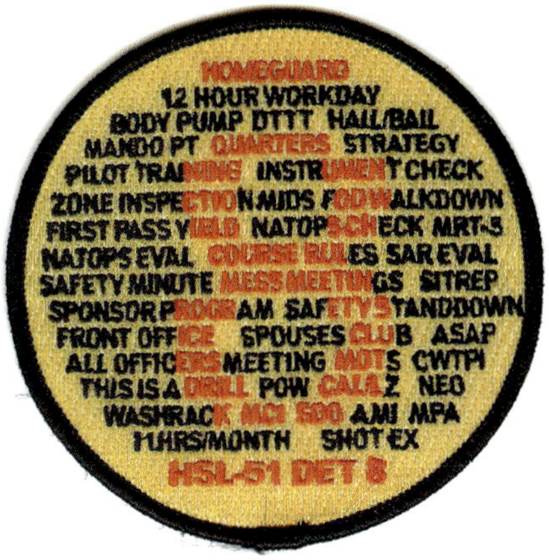

SL-51 Det. 6

HSL-51 Det. 7 Nightstalkers

HSL-51 Det. 8 Standard Brief

SL-51 Det. 6 USS *Mustin* (DDG-89) Quell the Storm, Ride the Thunder

HSL-51 Det. 7 Operation Enduring Freedom

HSL-51 Det. 8 Homeguard

SL-51 Det. 6 USS *Mustin* (DD-987) Shoulder to Shoulder

HSL-51 Det. 7 To Hell and Back

HSL-51 Det. 11 Commander Seventh Fleet

HSL-51 Det. 11 USS *Blue Ridge* (LCC-19) Commander Seventh Fleet (Subdued)

HSL-51 Det. 11 Black Beard (Subdued)

HSL-51 All Weapons Det. July 2002

HSL-51 Det. 11 Black Beard

HSL-51 Det. 11 USS *Anzio* (CG-68) USS *Enterprise* (CVN-65) Carrier Strike Group 2006

HSL-60 Jaguars

HSL-51 Det. 11 Black Beard 1993

HSL-51 Det. 11 Black Beard

HSL-60 Jaguars SH-60B Plane Captain

HSL-60 Jaguars Tactical AUF-Ness 2008–2009

HSL-60 Det. 1 USS *Underwood* (FFG-36) Have Guns Will Travel (Desert)

HSL-60 Det. 4 USS *Simpson* (FFG-56) MQ-8B Firescout

HSL-60 Det. 1 Dirty Birds

HSL-60 Det. 2 USS *Samuel B. Roberts* (FFG-58) Counter Narcotics Operations & AUF 2008–2009

HSL-74 Demon-Elves

HSL-60 Det. 1 USS *Underwood* (FFG-36) Have Guns Will Travel

HSL-60 Det. 3 USS *Elrod* (FFG-55) Night AUF January–July 2012 (Subdued)

HSL-84 Thunderbolts

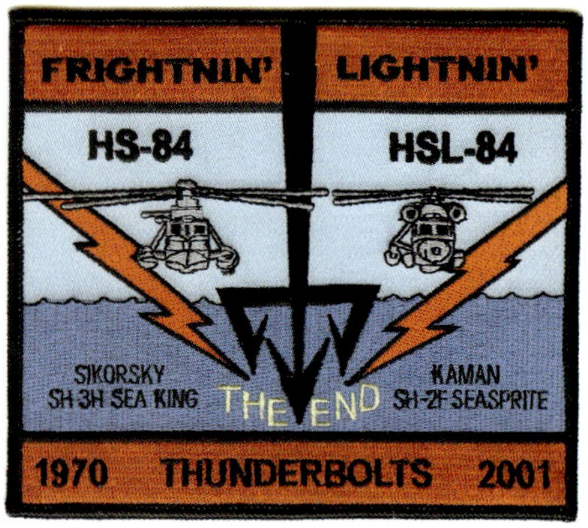

HSL-84 SH-2F SH-3H 1970–2001 The End

Unknown HSL Squadron Det. 21 USS *Horne* (CG-30)

HSM-35 Magicians

HSL-94 Titans

Multi-Squadron HSL-51 Det. 2 & HS-14 Det. "A" USS *Cowpens* (CG-63) Summer 2007

HSM-35 Magicians (Desert)

HSL-94 Det. 2 Bad to the Bone

HSM-31 Have Guns Will Hustle

HSM-35 Det. 7 NAS Fallon Det.

SM-35 Det. 7 Harbingers USS *Sampson* (DDG-10) BMD Cruise 2014–2015

HSM-37

HSM-41 Seahawks

SM-35 Det. 7 USS *Sampson* (DDG-10) Westpac 2014–2015

HSM-37 Det. 2 Nightmarchers USS *Howard* (DDG-83) Westpac 2014

HSM-41 Seahawks (Desert)

SM-35 Det. 8 USS *Lake Champlain* (CG-57) Rimpac 2014

HSM-40 Airwolves

HSM-41 Seahawk All Star

HSM-46 Det. 6 Big Irons USS *Vicksburg* (CG-69)

HSM-46 Det. 9 USS *Samuel B. Roberts* (FFG-58)

HSM-48 Vipers

HSM-48 Vipers (Subdued)

HSM-48 Vipers (Desert)

HSM-48 Vipers 25th Anniversary, 2014

HSM-48 Det. 1 USS *Leyte Gulf* (CG-55) Standard Navy Maritime Group February 2014

HSM-51 Warlords

HSM-51 Warlords

SM-51 I Saved the Day

HSM-70 Spartans

HSM-70 Carrier Air Wing 8

SM-51 Cherry Blossom Festival

HSM-70 Spartans (Subdued)

HSM-70 Det. 1 USS *Gettysburg* (CG-64) Persian Gulf 2011

SM-51 The Sun Rises on a New Era

HSM-70 Spartans (Desert)

HSM-70 Det. 3 USS *Philippine Sea* (CG-58) (Desert)

HSM-71 Raptors

HSM-71 Det. 2 Revolvers USS *Antietam* (CG-54)

HSM-71 Det. 2 USS *Stennis* (CVN-74) Strike Group Westpac 2009

HSM-71 Raptors (Desert)

HSM-71 Det. 2 Revolvers USS *Antietam* (CG-54) Westpac 2009

HSM-71 Det. 3 USS *Kidd* (DDG-100) Guardian of the Dark Sea

HSM-71 Westpac Navcent 2012–2013

HSM-71 Det. 2 USS *Stennis* (CVN-74) Strike Group Westpac 2009

HSM-71 Det. 9 Shogun

HSM-71 Det. 9 Shogun (Desert)

HSM-74 Swamp Fox

HSM-74 Swamp Fox (Desert)

HSM-72 Det. 1 Highlanders
USS *Hue City* (CG-66)

HSM-74 Swamp Fox

HSM-74 Det. 1 USS *Gravely*
(DDG-107) (Subdued)

HSM-73 Battlecats

HSM-74 Swamp Fox (Subdued)

HSM-74 Det. 1 USS *Gravely*
(DDG-107) (Desert)

HSM-75 Wolf Pack

HSL-75 Det. 4 USS *Princeton* (CG-59)
January–September 2013

HSM-77 Saberhawks (Subdued)

HSM-75 Wolfpack (Subdued)

HSM-77 Saberhawks

HSM-77 Saberhawks (Desert)

HSM-75 Wolf Pack (Desert)

HSM-77 Saberhawks

HSM-77 Det. 3 USS *Shoup* (DDG-86)
Bad Company

HSM-77 Det. 5

HSM-78 Blue Hawks

HSM-78 Blue Hawks MH-60R

HSM-77 Det. 5 (Subdued)

HSM-78 Blue Hawks (Subdued)

HSM-78 Blue Hawks MH-60R (Subdued)

HSM-77 Det. 5 (Desert)

HSM-78 Blue Hawks (Desert)

HSM-78 Blue Hawks MH-60R (Desert)

HSM-78 Det. 1 USS *Kidd* (DDG-100)
Westpac 2014

HSM-78 Det. 4 USS *Cape St. George* (CG-71)
Westpac 2014

HSM-79 Det. 4 USS *Cape St. George* (CG-71)
Westpac 2014 (Subdued)

HSM-79 Det. 4 USS *Cape St. George* (CG-71)
Westpac 2014 (Desert)

HT-8

HT-8

HT-8

HT-8 (Subdued)

HT-8 Det. NAS Key West, FL 2004

T-8 Det. NAS Key West, FL 2005

HT-18 CAT II Say All After Call Sign

HT-18 Instructor Pilot 1000 Hours

T-18

HT-18 CAT III Lima, Oscar, Bravo, Romeo, Bravo, Whatever

HT-18 Instructor Pilot 1500 Hours

T-18 Penguin No Flights- No X's- No Worries

HT-18 CAT IV Tension On, Easy Up

HT-28 Hellions

HT-28 Hellions (Subdued)

HU-1 Unit 23 USS *Burton Island* AGB-1 Arctic Dew Line Operation 1958

HU-2 Fleet Service

HT-28 Hellions (Desert)

HU-1 Det. Mike USS *Ranger* (CVA-61) Westpac 1965–1966

HU-2 Det. 62 Whirlydogs Med. 1963–1964

HU-1 Unit 6 USS *Rochester* (CA-124) Westpac 1957-1958

HU-2

HU-4

J-734 NAS Niagara Falls, NY

HUQ-1 Hydras

HX-21 Blackjack (Subdued)

J-871 NAS Oakland, CA

HX-1

HX-21 Blackjack

J-911 NAS Squantum, MA

HX-21 Blackjack

HX-21 Sixty-Five Years of Rotary Wing Flight Test

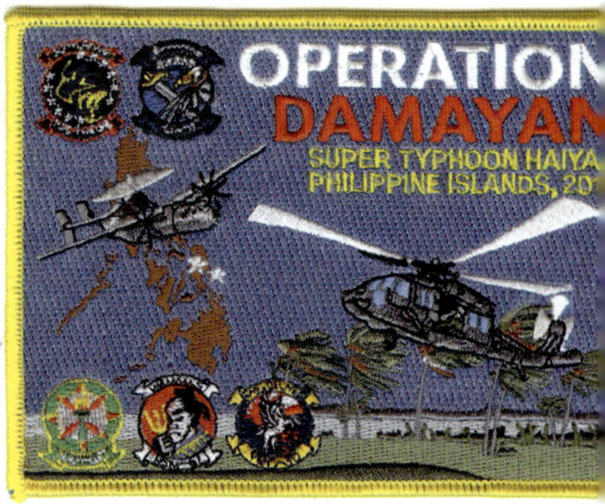

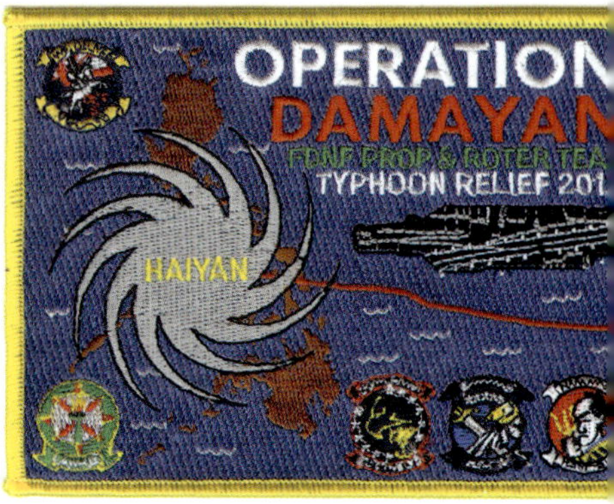

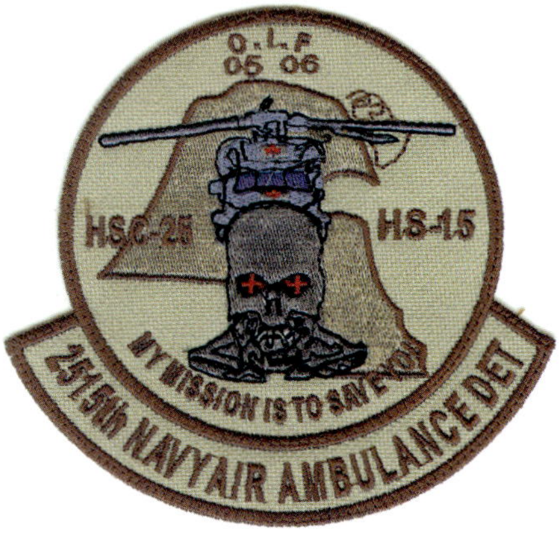

Multi-Squadron; HS-84 & HS-85 HMCS Provider (AOR 508)

Multi-Squadron; HSC-25 & HS-15-2515th Naval Air Ambulance Det. (Desert)

Multi-Squadron; HSC-25 & HS-15-2515th Naval Air Ambulance Det. Operation Iraqi Freedom 2005–2006 (Desert)

Multi-Squadron; HSC-25 & HS-15-2515th Naval Air Ambulance Det. Operation Iraqi Freedom 2005–2006 (Desert)

Multi-Squadron; HSL-43 Det. 4 & HSL-49 Det. 8 USS *McClusky* (FFG-41) 2012

Multi-Squadron; HS-14 & HSL-51 Det. 6B USS *Lassen* (DDG-82)

Multi-Squadron; HSC-12, HSM-51, HSM-77, VAW-115 & VRC-30 Operation Damayan Super Typhoon Haiyan Relief Philippine Islands 2013

Multi-Squadron; HSC-12, HSM-51, HSM-77, VAW-115 & VRC-30 Operation Damayan Super Typhoon Haiyan Relief Philippine Islands 2013

VXE-6 Antarctic Development Squadron Helo Crew

INDEX

GLOSSARY

C	Aircrew
MCM	Airborne Mine Counter Measures
SW	Anti-Submarine Warfare
SWOC	Anti-Submarine Warfare Operations Center
tl. / LANT	Atlantic Ocean
UF	Airborne use of Force
altops	Baltic Sea Operations
MD	Ballistic missile defense
enturian	100 Shipboard Landings
T OPS	Counter Illicit Trafficking
OOP	Continuity of Operations Plan
SAR	Combat Search and Rescue
TOC	Counter Transnational Organized Crime
et.	Detachment
E"	Battle Readiness Award
SG	Expeditionary Strike Group
DNF	Forward Deployed Naval Forces
RS	Fleet Replacement Squadron
ST	Helicopter Support Team
ex	North / South Pole Exercise
O.	Indian Ocean
SAR	Inverse synthetic aperture radar
OPA	Junior Officers Protection Association
W	Jungle Warfare
AMPS	Light Airborne Multi-Purpose System
SE	Landing Signal Enlisted
led.	Mediterranean Sea
IEF	Middle East Force
IIF	Maritime Interdiction Force
ISRT	Maritime Security Response Team
ATO	North Atlantic Treaty Organization
AVCENT	Navy Forces Central Command
ISRT	Maritime Security Response Team
Nocturian	100 Night Shipboard Landings
Norpac	Northern Pacific
NOSC	Navy Oceans Support Center
NVG	Night Vision Goggles
OEF	Operation Enduring Freedom
OIF	Operation Iraqi Freedom
OSW	Operation Southern Watch
Pac.	Pacific Ocean
Pacex	Pacific Exercise
Panamax	Panama Exercise
PC	Plane Captain
Rimpac	Pacific Rim
"S"	Safety Award
SAR	Search and Rescue
SNFL	Standard Naval Force Atlantic
SNFM	Standard Naval Force Mediterranean
SNMG	Standard Naval Maritime Group
The Ditch	Suez Canal
UAV	Unmanned Aerial Vehicle
UNITAS	Annual US - South American Allied Exercise
USNS	Naval Sealift Command Ship
Vertrep	Vertical Replenishment (Helicopter)
VOD	Vertical Onboard Delivery (Helicopter)
Westpac	Western Pacific

Unidentified Acronyms

BAMPG
BOD
CJTFEX
CNT
ISE
SDM
SHMG

AUTHOR'S BOOKS

JS Naval Aviation Patches; Carriers, Air Wings, Support Establishments, 1995, ISBN: 978-0-88740-753-6

JS Naval Aviation Patches; A/C, Attack Squadrons, Helicopter Squadrons, 1995, ISBN: 978-0-88740-801-4

JS Naval Aviation Patches; VF Squadrons, VFA Squadrons, VQ Squadrons, 1995, ISBN: 978-0-88740-802-1

9F-6, 6P Flight Manual (Reprint) 1995, ISBN: 978-0-88740-778-9

-4J Flight Manual (Reprint) 1995, ISBN: 978-0-88740-781-9

US Naval Patches; Commands, Supply, Amphibious, Seals, Fleets, Flotilla and Groups, 1996, ISBN: 978-0-7643-0068-4

US Naval Patches; Ships, Battle Ships, Cruisers, Destroyers, Tank Landing Ships, 1997, ISBN: 978-0-7643-0144-5

US Naval Patches; Submarines, 1997, ISBN: 978-0-7643-0186-5

Aircraft Histories of the LTV A-7 Corsair II (USN, Foreign) 2009, ISBN: 978-0-7643-3238-8

VC-8 Sikorsky SH-3H Sea King Landing on USS *Estocin* (FFG-15) during counter-narcotics training in the Caribbean Sea.